·80· GREAT COLLECTOR'S GARDEN PLANTS

·80· GREAT COLLECTOR'S GARDEN PLANTS

KEN DRUSE

CLARKSON POTTER/PUBLISHERS
NEW YORK

Published by Clarkson N. Potter, Inc., 201 East 50th Street, New York, New York 10022. Member of the Crown Publishing Group.

Random House, Inc., New York, Toronto, London, Sydney, Auckland
http://www.randomhouse.com/

Clarkson Potter, Potter, and colophon are trademarks of Clarkson N. Potter, Inc.

Printed in China

Design by Maggie Hinders and Lauren Monchik

Library of Congress Cataloging-in-Publication Data
Druse, Ken.
80 great collector's garden plants / Ken Druse.
I. Plants, Ornamental. 2. Plants, Ornamental—Pictorial works.
I. Title.
SB407.D77   1998   97-27047

ISBN 0-609-80084-1

10 9 8 7 6 5 4 3 2 1

First Edition

Acknowledgments

I want to thank some of the people who helped produce this book: Chip Gibson, President and Publisher of Crown Publishers; Lauren Shakely, Clarkson Potter's Editorial Director, who suggested this project and made the idea into reality; Maggie Hinders, the series designer, and Lauren Monchik, who designed this book; Mark McCauslin, production editor; and Joan Denman, production manager.

I also want to thank Laurie Stark, Joan De Mayo, Tina Constable, Wendy Schuman, and Mary Ellen Briggs for always doing more than their share on our projects. Thanks to Helen Pratt, my indefatigable agent; Ann Kearney-Dutton, photo librarian; Louis Bauer for his patience; and George Waffle for his friendship and efforts. I also must thank Ruth Clausen, whose help in making this book accurate was useful and invaluable.

Lastly, I have to thank you. As I travel and lecture around the country, I see new plants in gardens every day, and gardeners often tell me that my books have inspired them. It is gratifying to learn that my books have helped (and touched) so many.

Contents

Introduction

When I traveled around the country photographing native-plant gardens for my book *The Natural Habitat Garden*, I was impressed by how many people had created nearly wild areas along the outer edges of their properties, and smaller specialty gardens in beds or borders or snug up against the house. It was the resolution to the responsible gardener's dilemma: how to encourage indigenous plants while still acquiring wonderful botanicals from around the country and world. I chronicled these gardens of rare and unusual plants in my book *The Collector's Garden.*

The Collector's Garden also documents the people who seek these plants and the reasons why they do. Some collectors are *hunters*, who search the world, or even their own backyards, for plants new to horticulture. Others are *missionaries*, who promote an overlooked genus or a threatened plant and propagate them for home gardeners. *Specialists* zero in on one kind of plant or aspect of plants. For example, some gardeners are smitten by common plants that come in hundreds of variations, such as *Hemerocallis* (daylilies) and want to possess all the variations in flower form, color, and size. Other specialists collect plants that have a physical or morphological trait in common, such as cacti and succulents. Another kind of specialty collector seeks plants that come from a certain habitat type or plant community, such as plants that originate in alpine meadows.

The fourth type of collectors are the *aesthetes,* who seek plants simply because they find them irresistible. Most of us fall into this category. I want nearly every new plant I find beautiful, but with some restrictions. I do not collect invasive plants like running bamboo; rampant vines; or exotic berried plants that spread with the help of animals. I also do not "wild-collect"—steal plants from their original homes. The hunters in the book, for example, know that only a few seeds should be used to propagate new plants, and that cuttings or individual specimens can be removed only when a population is threatened by imminent development. Hundreds of plants in cultivation today were propagated from plants rescued from the paths of bulldozers.

The goal of this book is to give enough information

about these special plants so that you will know if it has a chance of being happy in your garden. This need not mean that you seek out only those plants that come from places similar to your own garden, but that you are sensitive to the plant's needs. For example, in my garden, I grow many frost-tender plants and store them indoors over winter, and I use herbaceous and woody plants from tropical and subtropical regions of the world for temporary color: ones such as canna, salvia, and coleus.

This book includes four sections: *Woody Plants,* like trees and shrubs; herbaceous ones, like *Hardy Perennials* and *Tender Plants and Annuals;* and *Climbing Plants* such as vines. Each plant is shown in a color photograph followed by cultural information, a pronunciation guide, descriptions of its attributes and uses, and best of all, examples of related species, cultivars, and companion plants. Altogether, there are hundreds of plants to collect and grow. The common thread, regardless of the type, is that all of these plants allow us to expand our palette and realize our wildest garden dreams and planting schemes. Collector's gardens are never dull, somber, or lethargic. They are rarely a solid expanse of green. We fill our plantings with brilliant leaves on shrubs, for instance, and sparkling flowers on vines.

Most of the plants featured in this book are just now

becoming available at specialty nurseries or through the mail. On the last pages of this book, you'll find a guide to mail-order sources for the plants illustrated and described. A few may not always be offered each year, but you can become a collector of catalogs, as well, and from time to time find nearly every plant you want to grow.

Another way to acquire rare plants is by joining a plant society. I highly recommend the North American Rock Garden Society and the Hosta Society, which have chapters all over the country. I encourage you to join the wildflower or native plant organization in your state, as well. You can meet wonderful people through these groups, attend symposia, classes, shows, and perhaps best of all, participate in plant swaps, seed exchanges, and plant sales. The names and addresses of many societies can be found in the library; *The Collector's Garden* has such a list. And if you are hooked up to the internet, just type in a search word.

Collecting is exciting and addictive. Stamp, antique car, coin, and even doll collectors know this. But unlike these people, the objects of our obsession are alive; and they can be increased. We can share our passion in a real way by making cuttings for new plants to inspire budding collectors. We don't have to be the only person on the block to grow a certain plant (maybe just the first).

Woody
Plants

Woody Plants

Herbaceous plants—those with soft tissues like the annuals and perennials—cannot be relied upon for a year-round framework. They come and go with the seasons and most have limited life spans. The woody plants, however—shrubs and trees—live for decades and, deciduous or evergreen, provide interest for more than just one season a year. Most have flowers, but all should have foliage characteristics. Rhododendrons, for instance, provide a solid mass in summer and green in the otherwise stark winter landscape—long after the showy spring flowers are gone.

A good example of a plant that "pays the rent," as some gardeners say, is *Hydrangea quercifolia,* the oakleaf hydrangea. This plant blooms with fragrant white flowers in early summer. Then the flowers deepen in color to pink, green, and finally walnut brown. In fall, the leaves turn burgundy bronze. In some years the leaves persist until spring; in others, the foliage drops to reveal papery curls of cinnamon-stick–like exfoliating bark.

I use woody plants more and more often to set the color schemes for my beds and borders, and use herbaceous ones to drape over the colorful structure and augment the arrangements. The colors come not only from flowers, but from the leaves of woody plants; the bark of trees, which often can be found in shades of silver, metallic red, or black; and from variegation—when leaves exhibit more than one color. The leaves of shrubs and trees can be found with cream, gold, chartreuse, pink, red, white, and other hues.

The current vogue is for shrubs with leaves in golden shades combined with ones that are red. For example, among elderberry shrubs alone, there are varieties with golden foliage (such as *Sambucus racemosa* 'Sutherland Gold') and others that are metallic maroon (*S. nigra* 'Purpurea', for instance). There are also elderberries that have green leaves splashed white (*S. nigra* 'Pulverulenta'), dark green ones edged gold (*S. nigra* 'Marginata'), or silver (*S. canadensis* 'Argenteo Marginata').

On the following pages, you will see many examples of woody plants with sensational flowers and often other colorful attributes. Only space will restrict your choices.

Acer palmatum cultivar (Japanese maple)

Aralia elata 'Variegata' (variegated Japanese Angelica-tree)

Berberis thunbergii 'Atropurpurea Nana' (Japanese barberry cultivar)

RIGHT
Brahea armata
(Mexican blue fan palm)

ACER PALMATUM CULTIVAR

PRONUNCIATION: AY-ser pal-MAY-tum
COMMON NAME: Japanese maple
HOMELAND: Of garden origin; species from
Japan, China, Korea
HARDINESS: USDA Zones 5–8
SIZE: 15'–20' tall and as wide, depending
upon the cultivar

INTEREST: Deeply fingered foliage with five or
more lobes, green to purplish-red in color,
turning brilliant red in fall
LIGHT CONDITIONS: Light to dappled shade
SOIL/MOISTURE: Well-drained, moisture-
retentive, fertile soil

DESCRIPTION: There are countless named cultivars of Japanese maples on the
market, with varying leaf shapes, color, and dissection. Some remain dwarf or
grow very slowly, rarely exceeding 6' or so in height. This small deciduous tree
blooms rather inconspicuously with umbels of small reddish-purple flowers in
early spring, before the leaves appear. Excellent as a specimen or accent plant,
or in a shrub collection or grouping.

ARALIA ELATA 'VARIEGATA'

PRONUNCIATION: a-RAY-lee-a ee-LAY-ta
COMMON NAME: Variegated Japanese
Angelica-tree
HOMELAND: Of garden origin; species from
Japan, China, eastern Russia
HARDINESS: USDA Zones 4–9
SIZE: 8'–10' tall; 8'–10' across

INTEREST: Terminal rosettes of compound,
cream-variegated leaves to 5' long, arranged
on thick, spiny stems; panicles of cream flowers in late summer.
LIGHT CONDITIONS: Sun to light shade
SOIL/MOISTURE: Fertile, well-drained soil

DESCRIPTION: This slow-growing deciduous tree suckers from the base and
forms large clumps or thickets. The very showy horizontal foliage makes it an
ideal specimen plant, but one that may be difficult to place in small gardens;
provide a dark background to show it off. Masses of tiny flowers borne at the
end of the stems produce a foamy effect which lasts several weeks. The flowers
are sometimes followed by blue-black berries which drop in the fall, if not con-
sumed by birds.

BERBERIS THUNBERGII 'ATROPURPUREA NANA'

PRONUNCIATION: BER-ber-is thun-BER-jee-a

COMMON NAME: Japanese barberry cultivar

HOMELAND: Of garden origin; species from Japan

HARDINESS: USDA Zones 4–8

SIZE: 1½'–2' tall; 2½'–3' across

INTEREST: Deciduous, dense, rounded shrub with reliably red to maroon small leaves on spiny stems; small yellow flowers in spring, red berries later.

LIGHT CONDITIONS: Full sun

SOIL/MOISTURE: Well-drained, even dry soils of average fertility

DESCRIPTION: This cultivar, 'Atropurpurea Nana', or 'Crimson Pygmy', as it is sometimes called, is one of the best choices for small-scale deep red foliage. It is important to get the true clone as there are many red-leaved cultivars and much nomenclatural confusion in the trade. Popular as an accent plant in mixed beds and borders, 'Atropurpurea Nana' is an ideal size for combining with perennials such as 'Autumn Joy' sedums, hardy geraniums, and coral bells. Try it with a skirt of *Ajuga reptans* 'Burgundy Glow'.

BRAHEA ARMATA

PRONUNCIATION: BRA-hee-a ar-MAH-ta

COMMON NAME: Mexican blue fan palm, blue hesper palm

HOMELAND: Mexico, Baha California

HARDINESS: USDA Zones 8–10

SIZE: 45' tall; top spread 6'–8'

INTEREST: Silver blue-green foliage, and showy, long inflorescences

LIGHT CONDITIONS: Full sun

SOIL/MOISTURE: Well-drained, average soil. Tolerates drought.

DESCRIPTION: Slow-growing Mexican blue fan palm has attractive silvery blue, waxy 3'–6'-wide leaves, divided into 40–50 segments; their petioles are armed with curved teeth. The leaves, atop stout trunks, move at the slightest breeze, creating a feeling of movement in the garden. The old leaves persist, unless cut or burnt off, forming a skirt around the trunk. The plumelike, arching inflorescence is 12'–15' long, and is composed of clusters of creamy white flowers, which are followed by ¾" fleshy yellow fruits, marked with brown.

Clematis ochroleuca (curly heads, leather flower)

Davidia involucrata (dove tree, handkerchief tree)

Enkianthus campanulatus (showy lanterns)

<small>LEFT</small>
Cornus controversa 'Variegata' (variegated pagoda dogwood)

CLEMATIS OCHROLEUCA

PRONUNCIATION: KLEM-a-tis o-kra-LOO-ka

COMMON NAME: Curly heads, leather flower

HOMELAND: Southern New York to Georgia

HARDINESS: USDA Zones 5–9

SIZE: 1'–2' tall; 1' across

INTEREST: Yellowish-white urn-shaped flowers flushed with purple in late spring and early summer; silky-haired, ovate foliage.

LIGHT CONDITIONS: Light to partial shade; full sun in cool summer regions.

SOIL/MOISTURE: Average, well-drained soil, enriched with humus

DESCRIPTION: The shrubby, nonvining clematis species, many of which are woody, are less well known than the climbing ones. Their flowers may be open as in *C. recta* and *C. heracleifolia*, or bell- or urn-shaped as in *C. integrifolia*, *C. scottii*, *C. freemontii*, and *C. ochroleuca*. They seldom top 3' in height, but some, especially *C. integrifolia*, are more scrambling than upright. Curly heads makes upright, rounded plants. The stems bear pairs of undivided, 4"-long, pointed leaves; the solitary 1" flowers are held in the upper leaf axils. The fluffy fruits that follow may reach 2" long. An interesting plant for a native plant or wild garden.

CORNUS CONTROVERSA 'VARIEGATA'

PRONUNCIATION: KOR-nus con-tro-VERS-a

COMMON NAME: Variegated pagoda dogwood, wedding cake tree

HOMELAND: Japan, China

HARDINESS: USDA Zones 5–8

SIZE: 30'–50' tall and as wide

INTEREST: Tiered branches bearing alternate, creamy white–bordered foliage; in late spring, clusters of white flowers with blue-black berries later.

LIGHT CONDITIONS: Light to partial shade

SOIL/MOISTURE: Moist, acid, well-drained soil

DESCRIPTION: This stunning specimen tree, grown for its elegant horizontal branching and airy variegated foliage, is best displayed against a background of dark woodland or evergreens. Its 3"–6" broadly elliptic leaves, irregularly edged with cream, remain handsome through the season, but will burn if sited in full sun. The young stems have reddish bark, but are not nearly as colorful as those of the red stem dogwood. The 3"–6"-wide clusters of flowers are attractive, and the berries that follow provide food for birds.

DAVIDIA INVOLUCRATA

PRONUNCIATION: dav-ID-ee-a in-vol-OO-kra-ta

COMMON NAME: Dove tree, handkerchief tree

HOMELAND: China

HARDINESS: USDA Zones 6–8

SIZE: 20'–40' tall and as wide

INTEREST: Broadly pyramidal in habit; in spring, two showy white bracts subtend each round, yellow flowerhead.

LIGHT CONDITIONS: Full sun to light shade

SOIL/MOISTURE: Moist, well-drained soil, high in organic matter

DESCRIPTION: This much-sought-after deciduous tree was introduced by E. H. Wilson. Although many trees grow in this country, they do not all flower reliably, even after a decade or more, and may become a disappointment. The striking bracts are of unequal length; the lower hangs down and may reach 7" long. Although they bloom for only about two weeks, their bright green foliage and attractive habit earn their keep through the summer. The scaly orange-brown bark adds winter interest. Use as a specimen plant, perhaps as a focal point at the end of a vista or allée, but leave sufficient space for the tree to show off its shape.

ENKIANTHUS CAMPANULATUS

PRONUNCIATION: en-ki-AN-thus kam-pan-u-LA-tus

COMMON NAME: Redvein enkianthus, showy lanterns

HOMELAND: Japan

HARDINESS: USDA Zones 5–9

SIZE: 8'–15' tall; 4'–6' across

INTEREST: Nodding clusters of red-veined cream bells in spring; brilliant red, orange, and yellow fall foliage.

LIGHT CONDITIONS: Partial shade

SOIL/MOISTURE: Moisture-retentive, highly organic acid soil

DESCRIPTION: Redvein enkianthus is one of the most beautiful shrubs to grow in light woodlands or in partly shaded positions in more open gardens. It tolerates city conditions well and also adapts readily to container culture. Underused in U.S. gardens, it deserves to be more popular, and is especially useful in limited space where its spectacular fall color can be appreciated as much as its spring display of bloom. An excellent companion for other lovers of acid soils such as rhododendrons and azaleas; carpet its feet with primulas, sweet woodruff, epimediums, or low ferns.

Hydrangea paniculata
(panicle hydrangea)

Ilex verticillata
(winterberry)

Juniperus chinensis 'Old
Gold' (Chinese juniper)

RIGHT
Lindera angustifolia

HYDRANGEA PANICULATA

PRONUNCIATION: hy-DRAN-jee-a
pan-ik-ewe-LA-ta
COMMON NAME: Panicle hydrangea
HOMELAND: Of garden origin; species from
Japan, China
HARDINESS: USDA Zones 3–8
SIZE: 10'-15' tall; 10'–15' across

INTEREST: In late summer to fall, large,
rounded, white flowerheads, which turn to
pink as they age; yellow fall color.
LIGHT CONDITIONS: Sun to partial shade
SOIL/MOISTURE: Moisture-retentive, fertile
soil

DESCRIPTION: The cultivar 'Paniculata Grandiflora', called peegee hydrangea, is ubiquitous, but look for the original species form, which has a more delicate flower panicle. Also seek out unusual varieties. A refined edition of the well-known peegee hydrangea, 'Tardiva' is more in scale with smaller gardens. Its flowerheads are somewhat smaller, to about 6″ long, with an abundance of slightly smaller sterile flowers. A good cut flower, for both fresh and dry use. All can be trained as a standard to punctuate the spine of mixed island beds; it also makes a late-season accent in shrub collections. Blooms on new wood, so prune very hard in early spring or after bloom time.

ILEX VERTICILLATA

PRONUNCIATION: EYE-lex ver-tiss-ill-AH-ta
COMMON NAME: Winterberry
HOMELAND: Nova Scotia west to Wisconsin
and south to Florida and Missouri
HARDINESS: USDA Zones 3–9
SIZE: 6'–15' tall or more and as wide

INTEREST: Clumps of gray-barked stems, with
deciduous dark green foliage. Female plants
bear ¼″ bright red berries in late summer.
LIGHT CONDITIONS: Full sun to partial shade
SOIL/MOISTURE: Moisture-retentive, acid, fertile to average soil

DESCRIPTION: This shrub has recently become popular as an ornamental, and many new cultivars and hybrids have been introduced. A male plant is required to pollinate the females for good fruit production. Some cultivars use specific pollinators. For example, 'Red Sprite' and 'Afterglow' are pollinated by 'Jim Dandy'; 'Southern Gentleman' (a. k. a. 'Early Male') pollinates 'Winter Red', 'Sunset', 'Jolly Red', and 'Sparkleberry'. Winterberry makes a fine specimen plant if kept pruned to shape, but is excellent massed in shaded, wet, or swampy areas, where it provides good cover and nesting sites for many species of birds. In some years the berries are stripped before New Year's, but in other seasons the birds leave them for later.

JUNIPERUS CHINENSIS 'OLD GOLD'

PRONUNCIATION: ju-NIP-er-us chi-NEN-sis

COMMON NAME: Chinese juniper

HOMELAND: Of garden origin; species from China, Japan, Mongolia

HARDINESS: USDA Zones 4–9

SIZE: 3' tall; 4' across

INTEREST: Bronze-gold foliage that retains its color through the winter

LIGHT CONDITIONS: Full sun

SOIL/MOISTURE: Well-drained, moisture-retentive soil of average fertility

DESCRIPTION: This dwarf, conical-shaped evergreen is most attractive when the foliage puts on new growth. Its golden yellow color gradually becomes bronzy. An excellent specimen plant, and compact enough for a place in large rock gardens, or as part of a collection of dwarf evergreens. Accent it with *Molinia caerulea* 'Variegata', give it a skirt of *Carex morrowii* 'Variegata', or contrast it with a dark backgound of yew or hemlock.

LINDERA ANGUSTIFOLIA

PRONUNCIATION: lin-DEER-a an-gus-ti-FO-lee-a

COMMON NAME: None

HOMELAND: Eastern Asia

HARDINESS: USDA Zones 6–8

SIZE: 10'–12' tall or more and almost as wide

INTEREST: Multistemmed, vase-shaped shrub with persistent foliage that turns orange-red and then apricot in fall; tan in winter. Blue-black berries.

LIGHT CONDITIONS: Full sun to very light shade

SOIL/MOISTURE: Average, well-drained soil

DESCRIPTION: This Asiatic spicebush is a welcome addition to the inventory of shrubs for American gardens, but is currently rare in the marketplace, due to its difficulty in propagation. Suckers gently. A striking backdrop for the white-bloomed stems of *Salix irrirata* and brilliant scarlet winter stems of *S. alba* 'Chermesena' as displayed at Wave Hill. Or, contrast it with the dark crimson fall color of *Rhus glabra* 'Laciniata'. Superb cut at the height of its fall color and placed in water, or cut later while in its apricot-tan dress and enjoyed as a long-lasting dried branch.

Magnolia × *wieseneri*
[*M. watsonii*]

Myrica pensylvanica
(northern bayberry)

Paeonia suffruticosa
(tree peony)

LEFT
Nandina domestica
(heavenly bamboo)

MAGNOLIA × WIESENERI [M. WATSONII]

PRONUNCIATION: mag-NO-lee-a × WIZ-ner-i
COMMON NAME: None
HOMELAND: Of garden origin: *M. hypoleuca* × *M. sieboldii*
HARDINESS: USDA Zones 5–8
SIZE: 15'–20' tall; 10'–15' across

INTEREST: In late spring, erect, fragrant, 6"–8" creamy-white, cup-shaped flowers accented with a cluster of crimson stamens
LIGHT CONDITIONS: Dappled to light shade
SOIL/MOISTURE: Deep, moist, but well-drained acid soil, enriched with humus

DESCRIPTION: This large-flowered, deciduous hybrid magnolia has leathery 4"–8" oval leaves which are undulate along the margins and accented with hairs along the veins. It develops a spreading crown shape with age and makes a fine multistemmed specimen. The flowers are goblet-shaped at first, opening wide after a few days; their spicy fragrance is very appealing. Blooms while quite young. Underplant with white-flowered narcissus such as 'Thalia', or 'Beersheba' with Siberian squills, and a ground cover of *Lamium macrophyllum* 'White Nancy'. Possibly hardier than reported.

MYRICA PENSYLVANICA

PRONUNCIATION: MI-rik-a pen-sil-VAN-ika
COMMON NAME: Northern bayberry
HOMELAND: Coastal communities from Nova Scotia to North Carolina
HARDINESS: USDA Zones 2–7
SIZE: 6'–10' tall and as wide

INTEREST: Aromatic, dark green leathery foliage, more or less evergreen; clusters of grayish-white berries in fall through winter.
LIGHT CONDITIONS: Full sun to partial shade
SOIL/MOISTURE: Adaptable in poor, sandy, or clay soils

DESCRIPTION: Northern bayberry suckers freely, forming large colonies that provide cover and nesting sites for birds. It is a fine plant for mass plantings, but is equally at home in shrub collections or in informal areas, especially in coastal gardens. Tolerates pollution well and is valuable in roadside plantings. Both male and female plants must be planted close together to get good crops of the white, waxy berries which are so attractive; these berries have been used to make bayberry candles since Colonial times. Seldom bothered by pests or diseases; prune only to shape, if necessary.

NANDINA DOMESTICA

PRONUNCIATION: **nan-DEE-na do-MES-tik-a**
COMMON NAME: **Heavenly bamboo**
HOMELAND: **China**
HARDINESS: **USDA Zones 6–9**
SIZE: **6'–8' tall; 2'–5' or more across**

INTEREST: **Evergreen compound leaves; large clusters of white 1/4" flowers in early summer and persistent red berries in fall.**
LIGHT CONDITIONS: **Full sun to shade**
SOIL/MOISTURE: **Moist, but well-drained, fertile soil**

DESCRIPTION: In spite of the connotations of its common name, heavenly bamboo is not invasive. Its leaves are reddish when young, turning green as the season progresses and then turning fiery red or crimson when the weather cools. Some cultivars color more brilliantly than others, but plants in full sun also color better. The clusters of 1/3" berries are very showy, and usually remain through the winter. To encourage dense, bushy growth, thin old stems annually. Group in a shrub collection, or use one of the smaller cultivars such as 'Atropurpurea Nana' as a rock garden specimen. Try the species or a tall cultivar such as 'Moyers Red' to screen unsightly structures. Trouble-free.

PAEONIA SUFFRUTICOSA

PRONUNCIATION: **PEE-o-nee suf-rut-ik-OSA**
COMMON NAME: **Tree peony, Mouton peony**
HOMELAND: **Northwestern China, Tibet**
HARDINESS: **USDA Zones 5–8**
SIZE: **4'–5' tall; 4'–5' across**

INTEREST: **Large single or double bowl-shaped flowers in late spring; deeply dissected, lobed, gray-green foliage.**
LIGHT CONDITIONS: **Light to partial shade**
SOIL/MOISTURE: **Rich, well-drained neutral or alkaline soil**

DESCRIPTION: Tree peonies have been cultivated in the Orient for centuries, but they have only recently become popular in this country. They have been hybridized extensively. There are hundreds of cultivars with sumptuous flowers in white and yellow, pinks, reds, and lavenders. The plants tend to be expensive since propagation is slow; many are grafted. Their attractive foliage is an asset in mixed or shrub borders, but give them enough space to show off.

Rhododendron vaseyi
(pinkshell azalea)

Rhododendron spp.
(rhododendron)

Rosa banksiae 'Lutea'
(Lady Banks' rose)

RIGHT
Robinia pseudoacacia 'Frisia'
(black locust)

RHODODENDRON SPP.

PRONUNCIATION: **ro-do-DEN-dron**

COMMON NAME: **Rhododendron**

HOMELAND: **Northern Hemisphere**

HARDINESS: **USDA Zones 4–9**

SIZE: **2'–15' tall and as wide or wider**

INTEREST: **In late spring, funnel- or bell-shaped flowers in every color blanket rounded bushes bearing leathery evergreen, semi-evergreen, or deciduous leaves**

LIGHT CONDITIONS: **Partial to light shade**

SOIL/MOISTURE: **Well-drained, moist, acid soil, enriched with humus**

DESCRIPTION: Rhododendrons are among the most popular shrubs. The leaves of evergreen rhododendrons are often felted beneath with white or brown fuzz or indumentum. Many species, including *R. maximum, R. yakusimanum,* and *R. catawbiense,* have been used extensively in hybridizing programs, resulting in a bewildering array of hybrids and cultivars to suit any site. Some of the best known include the evergreen white-, pink-, red-, or purple-flowered 'Kurume' series from Japan, which have found a place in many suburban American gardens. Also evergreen, the 'Girard' hybrids have large flowers and glossy leaves. The 'Dexter' hybrids are noteworthy for their dense foliage and flowers in a full array of colors. (The lavender-flowered hybrid 'Winterthur' is illustrated.)

RHODODENDRON VASEYI

PRONUNCIATION: **ro-do-DEN-dron VAY-see-eye**

COMMON NAME: **Pinkshell azalea**

HOMELAND: **Blue Ridge Mountains of North Carolina**

HARDINESS: **USDA Zones 4–8**

SIZE: **5'–10' tall; 3'–5' across**

INTEREST: **Clusters (trusses) of bell-shaped, pink flowers in late spring; the deciduous foliage emerges after bloom and sometimes displays red fall color.**

LIGHT CONDITIONS: **Full sun to light shade**

SOIL/MOISTURE: **Well-drained, moist, humusy soil**

DESCRIPTION: Pinkshell azalea is only one species of numerous deciduous rhododendrons, commonly referred to as azaleas. Although this species rarely hybridizes, others, including *R. viscosum* and *R. kaempferi,* have been been used extensively in the development of hybrids. The 'Exbury' and 'Knap Hill' hybrids, with off-white through yellow, orange, pink, and red flowers, are well known and have contributed in the parentage of popular cultivars and series. The 'Northern Lights' series were bred for hardiness and they thrive in Zone 4. Azaleas in a bewildering array of flower colors can be ordered from numerous mail order nurseries. Enthusiasts can also join the American Rhododendron Society.

ROBINIA PSEUDOACACIA 'FRISIA'

PRONUNCIATION: ro-BIN-ee-a
soo-do-a-KAY-se-a

COMMON NAME: Black locust, common locust

HOMELAND: Of garden origin; species from
Pennsylvania to Georgia and west to
Oklahoma and Missouri

HARDINESS: USDA Zones 4–9

SIZE: 30'–75' tall; 20'–35' across

INTEREST: Almost black, deeply furrowed bark;
compound, golden yellow foliage, and panicles of fragrant white pea flowers in spring.

LIGHT CONDITIONS: Full sun to light shade

SOIL/MOISTURE: Well-drained loamy soil, but
tolerates most

DESCRIPTION: An erect tree with an open crown which creates light, dappled shade. 'Frisia' has elegant yellow-green foliage, which usually retains its color; in some regions it may become green in summer and turn a brilliant golden in fall. The flowers are not as abundant as in the species, but the leaves more than earn the plant a place in the garden. Underplant with spring bulbs or a mass of foliage perennials such as epimediums, hostas (perhaps yellow-variegated culti-vars), lungworts, variegated Solomon's seal, or ferns. Susceptible to locust bor-ers, especially when stressed by drought; mulch at planting time. Tolerates all but severe drought and pollution when established. Seldom needs pruning.

ROSA BANKSIAE 'LUTEA'

PRONUNCIATION: RO-za BANKS-i-a

COMMON NAME: Lady Banks' rose

HOMELAND: Of garden origin; species from
China

HARDINESS: USDA Zones 6–9

SIZE: 15'–20' tall and as wide

INTEREST: Almost thornless, sprawling stems
covered with umbels of 1" yellow flowers in
late spring to early summer

LIGHT CONDITIONS: Full sun to partial shade

SOIL/MOISTURE: Average, well-drained soil

DESCRIPTION: Lady Banks' rose has dark foliage that remains evergreen in mild climates. The leaves are almost obscured by the slightly fragrant, buttery flow-ers during bloom time. This plant needs room to show off and may be difficult to keep within bounds. Excellent to clothe fences or to cascade over arbors or other structures. In cultivation in American gardens since the early 1800s, this hardy rose remains pest- and disease-free; it thrives on neglect. Valuable in coastal gardens for its salt tolerance.

Rosa glauca [*R. rubrifolia*]
(redleaf rose)

Rubus odoratus
(thimbleberry)

Sambucus nigra
'Pulverulenta'

Left
Rosa 'Mme. Alfred
Carrière' (rose)

ROSA GLAUCA [R. RUBRIFOLIA]

PRONUNCIATION: RO-za GLA-ka
COMMON NAME: Redleaf rose
HOMELAND: Central and southern Europe
HARDINESS: USDA Zones 2–8
SIZE: 5'–7' tall; 6'–9' across

INTEREST: Purplish-red glaucous foliage; single pink 1½" flowers and persistent red hips.
LIGHT CONDITIONS: Full sun to light shade
SOIL/MOISTURE: Well-drained, fertile soil

DESCRIPTION: This is one rose grown especially for its beautiful foliage. It tolerates more shade than other roses and is suitable for adding to lightly shaded shrub collections, which could include 'Pink Spires' sweet pepperbush for fall bloom. It also combines well with goat's rue, purple coneflowers, pink gladiolus, and other perennials and annuals in mixed borders. An interesting backdrop for pink Japanese anemones or the long, arching wands and purple flowers of *Lespedeza thunbergii.*

ROSA 'MME. ALFRED CARRIÈRE'

PRONUNCIATION: RO-za
COMMON NAME: Rose
HOMELAND: Of garden origin, parentage unknown
HARDINESS: USDA Zones 6–9
SIZE: 12'–18' tall

INTEREST: Clusters of fragrant, double white flowers in spring, repeating off and on through the season
LIGHT CONDITIONS: Full sun
SOIL/MOISTURE: Fertile, well-drained soil, enriched with organic matter

DESCRIPTION: This noisette-type climbing rose was introduced in France just over a century ago. Its strong canes have few thorns and bear semiglossy foliage that is moderately disease resistant. In cold regions, plant in a sheltered position and protect from freezing and thawing with a heavy winter mulch. Prune out dead and damaged wood in late winter, and thin the old canes to retain vigor. Deadhead routinely.

RUBUS ODORATUS

PRONUNCIATION: ROO-bus o-dor-AH-tus

COMMON NAME: Thimbleberry, flowering raspberry

HOMELAND: Nova Scotia to Georgia

HARDINESS: USDA Zones 2–8

SIZE: 6′–9′ tall; 6′ or so across

INTEREST: Maplelike foliage with purplish-pink, 2″ flowers in early summer. Red fruits appear in late summer.

LIGHT CONDITIONS: Light to partial shade

SOIL/MOISTURE: Well-drained, rich soil

DESCRIPTION: Thimbleberries make colonies of dense, but not thorny, deciduous shrubs, which provide cover, nesting sites, and food for birds. The fragrant flowers are arranged in clusters among the large leaves, which are white-felted beneath. Useful in a shrub collection or to attract wildlife.

SAMBUCUS NIGRA 'PULVERULENTA'

PRONUNCIATION: sam-BEW-kus NI-gra

COMMON NAME: European elderberry, common elderberry

HOMELAND: Of garden origin; species from Europe, northern Africa, southwest Asia

HARDINESS: USDA Zones 4–8

SIZE: 6′–8′ tall and as wide

INTEREST: Compoundly pinnate foliage, white at first, becoming green marbled with white; white flowers in spring, black berries later.

LIGHT CONDITIONS: Light dappled shade

SOIL/MOISTURE: Well-drained, moist, and fertile soil

DESCRIPTION: There are several cultivars of European elderberry on the market in the United States. Among the more readily available are the variegated 'Marginata' and 'Aureomarginata', which have leaves bordered with yellowish-white fading to cream and gold fading to yellow-green, respectively; 'Laciniata' has finely dissected green foliage. Showy but slow-growing 'Pulverulenta' is not presently available in this country, but is offered by several British nurseries. It makes a long-season foil for white-flowering rhododendrons, which serve as a contrasting foliar companion when not in bloom.

Syringa vulgaris 'Katherine Havemeyer' (lilac)

Viburnum lantana 'Variegata' (variegated wayfaring tree)

Weigela 'Rubidor'

RIGHT
Sambucus racemosa
'Plumosa Aurea'
(European red elder)

Sambucus racemosa 'Plumosa Aurea'

PRONUNCIATION: **sam-BEW-kus ray-see-MO-sa**

COMMON NAME: **European red elder**

HOMELAND: **Of garden origin; species from Europe, western Asia**

HARDINESS: **USDA Zones 3–7**

SIZE: **8′–12′ tall; 6′–8′ across**

INTEREST: **Compoundly pinnate leaves with finely cut leaflets, bright yellow maturing to green; yellowish-white flowers in spring; red berries.**

LIGHT CONDITIONS: **Full sun**

SOIL/MOISTURE: **Well-drained soil of average fertility**

DESCRIPTION: There are several attractive cultivars of the European red elder, mostly selected for their interesting foliage. The fernlike leaves of 'Tenuifolia' are finely dissected, on arching stems; more robust 'Sutherland Gold' also has cut leaves, golden when young, but not as fine as those of 'Plumosa Aurea'. All tolerate coppicing or cutting to ground level in early spring, which encourages the best foliage growth, on new wood; the flowers and fruits are forfeited for the leaves. The species has 3″–5″ oval panicles of pale green to yellowish-white flowers, followed by tightly packed clusters of small red berries in mid to late summer.

Syringa vulgaris 'Katherine Havemeyer'

PRONUNCIATION: **si-RING-ga vul-GAR-is**

COMMON NAME: **Lilac**

HOMELAND: **Of garden origin; species from southern Europe**

HARDINESS: **USDA Zones 3–8**

SIZE: **8′–15′ tall; 6′–10′ across**

INTEREST: **Heavy trusses of fragrant, double, lavender-pink flowers in late spring**

LIGHT CONDITIONS: **Full sun**

SOIL/MOISTURE: **Fertile, well-drained soil, enriched with humus**

DESCRIPTION: 'Katherine Havemeyer' is a popular old cultivar, introduced by the famous lilac hybridizing nursery of Lemoine in 1922. It is among the best of the pinks, although pink-flowered cultivars are susceptible to muddy or washed-out color changes according to the pH of the soil. Remove spent flowers, and prune out about one quarter of the older wood after bloom time, as well as weak or twiggy growth. The shrubs can be kept to a manageable height of 8′–9′; avoid legginess with routine pruning. An excellent cut flower. A dressing of lime is beneficial in acid soils.

VIBURNUM LANTANA 'VARIEGATA'

PRONUNCIATION: vi-BUR-num lan-TAN-a

COMMON NAME: Variegated wayfaring tree

HOMELAND: Of garden origin; species from Europe, western Asia

HARDINESS: USDA Zones 3–8

SIZE: 8'–10' tall; 8'–10' across

INTEREST: The felted, wrinkled leaves are woolly beneath and when young are splattered with yellow; white-variegated when mature.

LIGHT CONDITIONS: Light to partial shade

SOIL/MOISTURE: Well-drained, average soil

DESCRIPTION: This multistemmed shrub is currently scarce in the trade, although other cultivars of the wayfaring tree are more readily available. Compact 'Mohican' has handsome felted leaves, which emerge pale green at about the same time that the flat-topped clusters of creamy flowers appear. The leaves mature to dark green, a fine contrast for the showy fruits, which remain bright orange for several weeks before turning black.

WEIGELA 'RUBIDOR'

PRONUNCIATION: why-JELL-a

COMMON NAME: None

HOMELAND: Of hybrid origin

HARDINESS: USDA Zones 5–8

SIZE: 5'–6' tall; 5'–6' across

INTEREST: Bright gold, straw-colored foliage; in late spring, bright red funnel-shaped flowers.

LIGHT CONDITIONS: Full sun, but light shade in hot regions

SOIL/MOISTURE: Well-drained, fertile soil

DESCRIPTION: Many cultivars of *Weigela* are on the market, mostly with all green foliage and pink, white, or rosy flowers. All are easy to grow, but should be pruned regularly after bloom time to thin out old, dead, or damaged wood and encourage new growth. 'Rubidor' should be protected from intense afternoon sun to avoid leaf scorch or fading. 'Variegata' has pale yellow or cream leaf margins; compact 'Java Red' has purple foliage. Since *Weigelas* do not have attractive fruit and seldom display any fall color, those with interesting leaves after bloom time are valuable to carry their weight, especially where space is limited.

Hardy
Perennials

Hardy Perennials

Since 1990 or so, perennials have been the most popular plants for the garden, but while most gardeners stick to a fairly narrow repertoire of perennials, collectors discover the plants of tomorrow. Many of the ones that we seek will become standards in nurseries and garden centers in the next years. For now, we rely on specialty nurseries, mostly mail-order, and grow some from seeds. Some people are afraid of buying plants through the mail, or do not realize that there has been a quiet revolution in this industry over the last decade. Critics claim that only plants grown locally will perform well locally, but for the most part, that is not true. What's more, you simply cannot find the range of esoteric hardy perennials at the local garden center. Plants there will be larger, however, so press your local supplier to carry more unusual hardy perennials.

Years ago, the handful of mail-order nurseries shipping herbaceous perennials sent them dormant in a little wad of sphagnum peat moss wrapped in paper or plastic. Today, most companies will ask you to specify a date that you would like to receive your plants, or they estimate the shipping time based on your zip code. Plants rarely spend more than five days en route, most often in pots, and sometimes in full bloom—ready to go in the garden. However, if you suspect that a plant was grown in a greenhouse and the tissue seems a bit soft, a few days in a cold frame or in a shaded place out of the wind will help harden it off.

The real appeal of mail order is not convenience, but variety. There are quite literally thousands of herbaceous plants that can be ordered by mail. Receiving the catalogs in the winter and reading them is almost as much fun as receiving the plants in spring. Catalogs are also a significant source of knowledge. No longer the four-color commercial extravaganzas of the past, modern specialty nursery catalogs are rarely illustrated. Instead, the best ones pass the savings along to you in the size and quality of plants.

To find your favorite company, place several orders for hardy perennials from different sources at the same time. The only problem is that as these small companies get more popular—bigger—their plants may get smaller.

Armoracia rusticana
'Variegata' (variegated
horseradish)

Angelica gigas (Korean
angelica)

Aruncus dioicus 'Kneiffii'
(goat's beard)

RIGHT
Arisaema candidissimum

ANGELICA GIGAS

PRONUNCIATION: an-JEL-ik-a GUY-gas
COMMON NAME: Korean angelica
HOMELAND: Korea
HARDINESS: USDA Zones 5–7
SIZE: 5'–7' tall; 18"–24" across

INTEREST: Dusky fingered leaves on purple ribbed stems; in late summer, purple buds explode into 6"–8" umbels of tiny purple flowers.
LIGHT CONDITIONS: Full sun to partial shade
SOIL/MOISTURE: Average, well-drained soil

DESCRIPTION: Angelicas are not for the faint-hearted gardener. This species makes a dramatic architectural accent in any design. Try it as a hedge between a perennial border and a background fence, or use it to punctuate a silver or white design. In a sunny spot, let it grow through shrubby *Artemisia* 'Powis Castle' or echo the deep crimson of *Lilium* 'Black Beauty'; partner it with bold hostas in shade. Valuable in floral arrangements, both in flower and in fruit. It often behaves as a biennial. Unless deadheaded, angelica seeds about freely; it is easy to have a good supply of seedlings on hand.

ARISAEMA CANDIDISSIMUM

PRONUNCIATION: ar-is-EE-ma kan-did-ISS-im-um
COMMON NAME: None
HOMELAND: Western China
HARDINESS: USDA Zones 7–9
SIZE: 12"–15" tall; 10"–12" across

INTEREST: In early summmer, elegant 6" white spathes, striped with green and pink inside, each sheathing a gray-green spadix
LIGHT CONDITIONS: Light shade
SOIL/MOISTURE: Fertile, moisture-retentive soil, enriched with humus

DESCRIPTION: Exotic-looking in bloom, this is an interesting and unusual plant for shaded gardens. The waxy spathe is delicately marked, flaring at the mouth, and with undulating edges. This aroid comes up in late spring; mark its position in the garden to avoid inadvertent damage. Best where it can be sheltered from drying winds; protect from slug and snail damage. Colonizes slowly, making large stands.

ARMORACIA RUSTICANA 'VARIEGATA'

PRONUNCIATION: ar-mor-AY-see-a
rust-ik-AN-a

COMMON NAME: Variegated horseradish

HOMELAND: Of garden origin; species from
southeastern Europe

HARDINESS: USDA Zones 3–9

SIZE: 2'–3' tall; 1'–2' across

INTEREST: Bold 15" leaves splashed and marbled with creamy white

LIGHT CONDITIONS: Full sun

SOIL/MOISTURE: Deep, moist, fertile soil

DESCRIPTION: Variegated horseradish has deep, thick, pungent roots which can be used for the same culinary purposes as the species. The oblong leaves have undulating margins and are sometimes irregularly lobed. The panicles of tiny white flowers are insignificant. Do not allow to spread uncontrolled.

ARUNCUS DIOICUS 'KNEIFFII'

PRONUNCIATION: ah-RUN-kus die-OH-ik-us

COMMON NAME: Goat's beard

HOMELAND: Of garden origin; species from
the Northern Hemisphere

HARDINESS: USDA Zones 3–8

SIZE: 3'–4' tall; 3'–4' across

INTEREST: Dark green, threadlike, compound
2'–3' leaves; large, fluffy panicle of tiny
white flowers in late spring to summer.

LIGHT CONDITIONS: Full sun to light shade

SOIL/MOISTURE: Moisture-retentive and fertile
soil

DESCRIPTION: This cultivar is worth growing for its delicately cut leaves, which give an airy effect. The plant is also much smaller than the species, which may attain 7' in height. Goat's beard is a plant of shrublike proportions, reminiscent of an enormous astilbe. Its leaves are tri-pinnately divided, each oval leaflet serrated. The handsome foliage of both 'Kneiffii' and the species are excellent as a foil for other flowers. Give the species room to show off. Male and female flowers are on separate plants; the males are more showy but it is unlikely they will be sexed in the nursery. Low maintenance.

Astrantia major
(masterwort)

Campanula portenschlaagiana
(Dalmatian bellflower)

Carex elata 'Aurea' (Bowles'
golden sedge)

Left
Asarum shuttleworthii
(mottled wild ginger)

ASARUM SHUTTLEWORTHII

PRONUNCIATION: ay-SAIR-um shut-l-WORTH-ee-i

COMMON NAME: Mottled wild ginger

HOMELAND: Virginia and West Virginia to Georgia and Alabama, in upland woods

HARDINESS: USDA Zones 7–9

SIZE: 6"–8" tall; 10"–12" across

INTEREST: Evergreen, 2"–4" heart-shaped leaves, solid green or mottled with silver. Almost hidden, vase-shaped, brown 2" flowers in spring.

LIGHT CONDITIONS: Bright or partial shade to shade

SOIL/MOISTURE: Moist or dry woodland soil

DESCRIPTION: This fine native makes an attractive foliage ground cover under trees and in woodland gardens, but should be planted close as it colonizes slowly. The curious flowers, which are spotted with violet on the inside, are pollinated by tiny flies attracted by a smell reminiscent of rotting meat. 'Callaway' is a selection with smaller leaves with prominent silver marbling. Prone to slug damage, especially in damp soils. Other North American species include deciduous *A. canadense,* and evergreen *A. virginicum* and *A. caudatum,* a western species.

ASTRANTIA MAJOR

PRONUNCIATION: as-TRAN-tee-a MAY-jor

COMMON NAME: Masterwort

HOMELAND: Europe

HARDINESS: USDA Zones 4–7

SIZE: 2'–3' tall; 18"–24" across

INTEREST: Mounds of deeply lobed, rounded leaves; domed umbels of greenish or pink flowers, with purplish bracts in early summer.

LIGHT CONDITIONS: Light to partial shade

SOIL/MOISTURE: Moisture-retentive soil, enriched with humus

DESCRIPTION: Masterwort is a valuable perennial for shade gardens, where its quiet charm shines. The foliage remains good-looking all season. The variegated cultivar 'Sunningdale Variegated' is particularly handsome. There are several notable cultivars selected for their colorful bracts: 'Alba' is all white; 'Rubra' has a collar of rosy pink bracts; 'Shaggy', sometimes listed as 'Involucrata', has irregularly long, pink bracts. A good cut flower.

CAMPANULA PORTENSCHLAAGIANA

PRONUNCIATION: **cam-PAN-ewe-la port-en-schlag-ee-AH-na**

COMMON NAME: **Dalmation bellflower**

HOMELAND: **Southern Europe**

HARDINESS: **USDA Zones 4–9**

SIZE: **5"–9" tall; 10"–12" across**

INTEREST: **Mats of dark green foliage covered with panicles of 1" purplish-blue flowers in late spring**

LIGHT CONDITIONS: **Full sun to very light shade**

SOIL/MOISTURE: **Excellent drainage, average soil**

DESCRIPTION: This very free-blooming perennial is ideal for planting in rock walls or tumbling over them. It also makes a showy ground cover at the front of mixed or perennial plantings. The long-petioled oval or heart-shaped leaves remain attractive after bloom time. Shear after flowering for neatness. An ideal companion for *Iberis sempervirens* or *Artemisia* 'Silver Brocade'.

CAREX ELATA 'AUREA'

PRONUNCIATION: **KAR-ex ee-LAY-ta**

COMMON NAME: **Bowles' golden sedge**

HOMELAND: **Of garden origin; species from northern Europe**

HARDINESS: **USDA Zones 5–9**

SIZE: **18"–24" tall and as wide**

INTEREST: **Semi-evergreen clumps of bright yellow grassy foliage, up to $\frac{1}{2}$" wide; narrow brownish spikes in early summer.**

LIGHT CONDITIONS: **Part sun to light shade**

SOIL/MOISTURE: **Constantly moist, acid soil**

DESCRIPTION: Bowles' golden sedge brightens damp, shaded areas of the garden in spring when the new growth emerges, but, unfortunately, the color fades as the season progresses. The flower spikes, while not showy, are held on upright or arching stems. The plant is at its best massed at the water's edge, or planted as a specimen. Protect from drying winds in hot areas; it prefers a humid atmosphere.

Filipendula ulmaria 'Aurea'
(golden meadowsweet)

Helleborus orientalis
(lenten rose)

Hemerocallis 'Joan Senior'
(daylily)

RIGHT *Corydalis flexuosa*
'Blue Panda' (blue panda
corydalis)

CORYDALIS FLEXUOSA 'BLUE PANDA'

PRONUNCIATION: kor-ID-al-is FLEX-oo-os-a

COMMON NAME: Blue Panda corydalis

HOMELAND: Of garden origin; species from western Sichuan province of China

HARDINESS: USDA Zones 5–7

SIZE: 12"–15" tall; 12" or so across

INTEREST: Attractive clumps of ferny foliage; electric blue, spurred flowers, $\frac{1}{2}$"–$\frac{3}{4}$" long from spring through fall.

LIGHT CONDITIONS: Partial shade

SOIL/MOISTURE: Moisture-retentive soil, enriched with humus

DESCRIPTION: This recent introduction has taken the plant world by storm. The flowers are a difficult color to use in the garden. It may be best planted among non-flowering maidenhair or other ferns; *Geranium macrorrhizum* 'Ingwerson's Variety' is another good companion. The plants should not dry out during the growing season, and they will rot if drainage is poor during prolonged freezing weather. Sometimes tricky to bring through the winter.

FILIPENDULA ULMARIA 'AUREA'

PRONUNCIATION: fil-ip-END-ewe-la ul-MARE-ee-a

COMMON NAME: Golden meadowsweet

HOMELAND: Of garden origin; species from western Asia, Europe

HARDINESS: USDA Zones 3–9

SIZE: 3'–6' tall; 2'–3' across

INTEREST: Bold, pinnately compound foliage, golden-yellow when young, later becoming cream

LIGHT CONDITIONS: Light shade

SOIL/MOISTURE: Moist, fertile soil

DESCRIPTION: Golden meadowsweet is grown for its attractive foliage and is best if not allowed to flower. Any resulting seedlings will probably have green leaves; propagate vegetatively from good color selections. The terminal leaflet is enlarged with 3–4 pairs of lateral leaflets. In warm climates, the foliage will bleach out badly in sun, but in the north a sunny spot is acceptible. A good companion for blue columbines, larkspur, and Siberian iris.

HEMEROCALLIS 'JOAN SENIOR'

PRONUNCIATION: hem-er-o-KAL-is

COMMON NAME: Daylily

HOMELAND: Hybrid of garden origin

HARDINESS: USDA Zones 3–10

SIZE: 18"–24" tall and as wide

INTEREST: Grassy foliage with strong, erect scapes bearing rounded, ruffled, 6" pristine white flowers, green at the throat

LIGHT CONDITIONS: Full sun to light shade

SOIL/MOISTURE: Well-drained, deep, fertile soil

DESCRIPTION: 'Joan Senior' is possibly the very best and most popular of the white daylily cultivars. It has won numerous awards and is considered the standard for its color. 'Joan Senior' blooms at the middle of the daylily season and repeats later. Daylilies, in general, are easy to grow and require little care beyond regular deadheading to remove unsightly, spent flowers, and dividing from time to time. Excellent for controlling erosion on difficult banks, in mixed or perennial gardens, and as specimens or container plants. Countless cultivars are on the market in a wide range of colors, except blue. Daylily enthusiasts can join the American Hemerocallis Society, which has chapters in various parts of the country.

HELLEBORUS ORIENTALIS

PRONUNCIATION: hell-e-BOR-us o-ree-en-TAL-is

COMMON NAME: Lenten rose

HOMELAND: Greece, Asia Minor

HARDINESS: USDA Zones 4–9

SIZE: 15"–18" tall; 12"–15" across

INTEREST: Evergreen, broadly fingered leaves and clusters of 2"–4" white to maroon flowers in early spring

LIGHT CONDITIONS: Medium shade

SOIL/MOISTURE: Highly organic, well-drained soil, enriched with compost or leaf mold

DESCRIPTION: Not as temperamental as its close relative, the Christmas rose (*H. niger*), Lenten roses bloom a little later and have several bowl-shaped flowers atop each 12"–18" stem. They are often copiously speckled with maroon at the center. After the petals fall, the calyx persists for several weeks, providing interest. The clean, bold foliage serves as an attractive contrast for delicate ferns, astilbes, and goat's beard in shaded or woodland settings. Try echoing the leaf shape with an underplanting of sweet woodruff. Valuable as an underplanting for fine-leaved trees and shrubs, such as honey locust, mountain ash, Japanese maples, and *Neillia sinensis*. A long-lasting cut flower.

Heuchera 'Persian Carpet'
(coral bells)

Hosta 'Frances Williams'
(plantain lily)

Iris douglasiana

LEFT
Miscanthus sinensis
'Purpurascens'
(flame grass)

HEUCHERA 'PERSIAN CARPET'

PRONUNCIATION: **HOO-ka-ra**

COMMON NAME: **Coral bells, alum root**

HOMELAND: **Of garden origin**

HARDINESS: **USDA Zones 5–7**

SIZE: **12″–18″ tall and as wide**

INTEREST: **Evergreen, rounded and lobed deep purple foliage, strongly marked with silver, wavy along the margins**

LIGHT CONDITIONS: **Full sun to light or partial shade**

SOIL/MOISTURE: **Well-drained soil, enriched with humus**

DESCRIPTION: Coral bells are currently riding the crest of popularity with plantsmen and recently have gained recognition as a foliage plant. Cultivars with purple, silver, marbled, mottled, and variegated foliage are available. 'Persian Carpet' is one of the large number of *Heuchera* introductions from Dan Heims in Portland, Oregon; it is probably a hybrid. Formerly, this genus was grown chiefly for its attractive and delicate spikes of pink, red, or white flowers in spring, and most were listed as cultivars of *H.* × *brizoides.* All cultivars are ideal for sunny or shaded gardens; in beds and borders; as a specimen, massed as a ground cover; or as a companion for native woodland plants. Dress with a skirt of silvery lamb's ears in sun, or complement with foamflower in shade.

HOSTA 'FRANCES WILLIAMS'

PRONUNCIATION: **HOS-ta**

COMMON NAME: **Plantain lily, funkia**

HOMELAND: **Of garden origin; species from Japan**

HARDINESS: **USDA Zones 3–9**

SIZE: **2½′–3′ tall; to 4′ across**

INTEREST: **Bold clumps of large, cupped, and crinkled blue-green leaves, irregularly edged with gold or cream; spikes of almost white flowers in summer.**

LIGHT CONDITIONS: **Partial to dappled shade**

SOIL/MOISTURE: **Well-drained, moisture-retentive soil**

DESCRIPTION: 'Frances Williams', an *H. sieboldiana* type, is among the most popular of the hundreds of hosta cultivars on the market. It is spectacular as a specimen, as a punctuation point where shady paths intersect, or in containers. Hosta may be deep green, blue-green, blue, yellow, yellow-green, or variegated with white or yellow, often of "good substance," and puckered or strongly veined. The leaves vary from 1″–2″ long to 2′ in length. A few are grown for their spikes of lily-like, sometimes fragrant white or lavender flowers. Mass large- or small-leaved cultivars as a ground cover; use low-growing ones to edge the front of a border. Enthusiasts can join the American Hosta Society.

Iris douglasiana

PRONUNCIATION: EYE-ris doo-glas-ee-AN-a
COMMON NAME: None
HOMELAND: Coastal regions from southern California to southwestern Oregon
HARDINESS: USDA Zones 6–10
SIZE: 18"–30" tall; 1'–2' across
INTEREST: Dark evergreen foliage, strongly veined; branched stems each with 2 or 3 pale cream to lavender and deep red-purple flowers in spring.
LIGHT CONDITIONS: Full sun
SOIL/MOISTURE: Lime-free, well-drained, moisture-retentive soil

DESCRIPTION: This iris was named for David Douglas, an early botanical explorer on the West Coast. It hybridizes freely with other iris in the Pacific Coast iris group, which also includes *I. innominata,* the other parent for the Pacific Coast hybrids. The flowers are beardless, usually in shades of lavender, and are strongly veined with purple. They form large clumps and are easy to grow, as they tolerate most soils; they even tolerate drought, although this is far from ideal. If growing from seed, select good color forms as some are wishy-washy.

Miscanthus sinensis 'Purpurascens'

PRONUNCIATION: mis-KAN-thus sy-NEN-sis
COMMON NAME: Flame grass, purple silver grass
HOMELAND: Of garden origin; species from eastern Asia
HARDINESS: USDA Zones 6–9
SIZE: 4'–6' tall; 3'–5' across
INTEREST: Compact clumps of ¼"–½"-wide grassy leaves; silvery flower plumes in mid-summer. Red-orange to red-brown fall color.
LIGHT CONDITIONS: Full sun to light shade
SOIL/MOISTURE: Moist, well-drained, fertile soil

DESCRIPTION: Flame grass colors best in cool areas. Its showy silver plumes of flowers contrast brilliantly with the mass of red-orange foliage below. Excellent combined with 'Autumn Joy' sedums and Korean chrysanthemums in perennial gardens, with other grasses and wildflowers such as black-eyed Susans in open meadows, or as contrast with woody plants in shrub collections.

Nelumbo nucifera 'The Queen' (sacred lotus)

Onosma alboroseum

Phlox subulata 'Tamanongalei'

Nelumbo nucifera 'The Queen'

PRONUNCIATION: nel-UM-bo nu-SIF-er-a

COMMON NAME: Sacred lotus, sacred lily, East Indian lily

HOMELAND: Tropical and subtropical Southeast Asia to Australia

HARDINESS: USDA Zones 10–11
Size: 1'–5' tall; 6'–8' across

INTEREST: Mostly floating, almost round, waxy leaves with solitary, fragrant, peonylike flowers held out of the water, in summer

LIGHT CONDITIONS: Full sun to very light shade

SOIL/MOISTURE: Sandy or average soil, beneath up to 1' of water

DESCRIPTION: The many-petaled, fragrant flowers of sacred lotus may reach 10" across. After blooming, the circular pods containing the seeds remain decorative and are often cut for dried arrangements. The leaves, which stand 1'–3' out of the water, are 1'–3' across, often with wavy margins. The extensive rhizomes can be cut into pieces to increase stock after growth has begun in spring. In warm climates where the water temperature does not fall below 40–45°F, and where the rhizomes can be covered by 3' of water to insulate them, the plants can remain outdoors all year-round. Otherwise, they must be brought under cover until water reaches 65°F the following spring.

Nymphaea odorata (TROPICAL)

PRONUNCIATION: nim-FAY-ee-a o-dor-AH-ta

COMMON NAME: Fragrant water lily, white water lily

HOMELAND: Eastern North America and Mexico

HARDINESS: USDA Zones 10–11
SIZE: 3"–12" tall; 4'–7' across

INTEREST: From summer through fall, fragrant, starry, 3"–5" flowers, with a central cluster of yellow stamens; floating, circular leaves.

LIGHT CONDITIONS: Full sun to very light shade

SOIL/MOISTURE: Rich clayey soil, enriched with manure, under 12"–18" of water

DESCRIPTION: Tropical water lilies are among the most dramatic of flowers. They range in color from white and yellows to pinks and reds, with several hybrid cultivars in purples and blues. Some, such as 'Red Flare' and pink 'Mrs. George C. Hitchcock', open at night; others, including 'Wood's Blue Goddess' and 'Yellow Dazzler' are day-bloomers. Most are fragrant. The flowers are held just above the water surface, but the green or brown foliage, often mottled or streaked with maroon, is usually floating. In cold climates, delay planting out until the water temperature has reached 70°F.

ONOSMA ALBOROSEUM

PRONUNCIATION: on-OZ-ma al-bo-RO-zee-um
COMMON NAME: None
HOMELAND: Mediterranean region, east to the Himalayas
HARDINESS: USDA Zones 7–9
SIZE: 8"–10" tall; 6"–8" across

INTEREST: Fuzzy gray-green leaves; upright stems with cymes of tubular, white flowers that turn reddish-purple and violet.
LIGHT CONDITIONS: Full sun
SOIL/MOISTURE: Very well-drained, dry soil

DESCRIPTION: Excellent in rock gardens where winter drainage is excellent, *Onosma* makes rounded clumps of bristly grayish foliage (which may cause skin irritation for some gardeners). The long-lasting 1" flowers open white or purplish-pink and deepen in color as they mature. They are arranged in gracefully drooping scorpoid cymes. An unusual ground cover for dry, sunny areas.

PHLOX SUBULATA 'TAMANONGALEI'

PRONUNCIATION: FLOX sub-ewe-LAH-ta
COMMON NAME: Moss pink, mountain phlox
HOMELAND: Of garden origin; species from New York to Maryland and Michigan
HARDINESS: USDA Zones 3–8
SIZE: 6"–9" tall; 12"–18" across

INTEREST: Mats of evergreen, needled foliage, covered in early spring with white flowers candy-striped with cerise
LIGHT CONDITIONS: Full sun
SOIL/MOISTURE: Well-drained soil of average fertility

DESCRIPTION: This Tasmanian selection is among the showiest of the moss pinks. All are fine rock garden plants and make excellent ground covers for sunny sites. Shear after bloom time for neatness and to expose the mats of dark green foliage. There is a large number of cultivars on the market with flowers in lavender-blue, pinks, rose, and white, sometimes with a contrasting eye.

Pulmonaria saccharata
'Margery Fish' (lungwort)

Pulsatilla vulgaris
[*Anemone pulsatilla*]
(Pasque flower)

Rudbeckia maxima
(giant coneflower)

LEFT
Tovara virginiana 'Variegata'
[*Polygonum virginianum*
'Variegatum'] (Virginia
tovara)

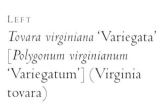

PULMONARIA SACCHARATA 'MARGERY FISH'

PRONUNCIATION: pul-mon-AIR-ee-a
sak-ar-AH-ta

COMMON NAME: Lungwort, Bethlehem sage

HOMELAND: Of garden origin; species from
southeastern France, Italy

HARDINESS: USDA Zones 3–8

SIZE: 12"–18" tall; 18"–24" across

INTEREST: Clumps of silver-spotted leaves,
which expand after bloom time; in spring,
pink flowers which turn bright blue.

LIGHT CONDITIONS: Partial to light shade

SOIL/MOISTURE: Well-drained, moisture-
retentive, humusy soil

DESCRIPTION: Lungworts are excellent planted closely as a ground cover in shade, where they contrast well with ferns, as well as with the fingered, dark foliage of hellebores. The leaves may expand to 1' or so long. 'Margery Fish', named for the noted English garden maker, has long been grown, and is often used as a parent for the newer cultivars. Many of these, including 'Excaliber', 'Spilled Milk', and 'Berries and Cream', make spectacular specimen plants, and are at their best where the distinctive variegated foliage can show off without competition. Susceptible to mildew where soil becomes dry.

PULSATILLA VULGARIS [ANEMONE PULSATILLA]

PRONUNCIATION: pul-sat-ILL-a vul-GAR-is

COMMON NAME: Pasque flower

HOMELAND: Northern Europe

HARDINESS: USDA Zones 5–8

SIZE: 12"–15" tall and as wide

INTEREST: Bell-shaped purple flowers in early
spring, before the silky fingered foliage fully
develops; feathery seed heads follow.

LIGHT CONDITIONS: Full sun to partial shade

SOIL/MOISTURE: Very well-drained neutral to
alkaline soil

DESCRIPTION: Pasque flowers bloom in a range of colors from white and pale lavender to pinks, rust, and deep purple; new selections from Europe expand the color range. The flowers, once a source of dye for coloring Easter eggs, are effective in rock gardens and make good companions for early spring bulbs such as squills and small narcissus species. Protect from intense sun in southern regions. The soil must drain freely to avoid crown rot during wet winters; add coarse grit to the soil and as a top dressing if drainage is questionable.

Rudbeckia maxima

PRONUNCIATION: **rud-BEK-ee-a MAKS-im-a**
COMMON NAME: **Giant coneflower**
HOMELAND: **Missouri to Mississippi, Louisiana, and Texas**
HARDINESS: **USDA Zones 6–9**
SIZE: **4'–10' tall; 2' or more across**

INTEREST: **Clumps of silvery blue-gray leaves; tall, sparsely-branched stems topped with black-eyed Susan flowerheads in summer.**
LIGHT CONDITIONS: **Full sun**
SOIL/MOISTURE: **Moist soil of average fertility**

DESCRIPTION: This species of black-eyed Susan is grown for its attractive leaves. These are undivided, smooth with a whitish bloom, and may reach 1' or so long. The stem leaves are slightly cupped and clasp the stem. The 2"–3" flowerheads have gold, drooping ray flowers and an elongated black disk. Valuable in informal gardens, especially in seaside locations, where they make good companions for daylilies and ornamental grasses.

Tovara virginiana 'Variegata'

PRONUNCIATION: **toe-VAR-a vir-jin-ee-AYE-na**
COMMON NAME: **Virginia tovara**
HOMELAND: **Eastern North America, Japan, Himalayas region**
HARDINESS: **USDA Zones 5–8**
SIZE: **2'–4' tall and as wide**

INTEREST: **Clumps of elliptical leaves, irregularly splashed with creamy white, marked with maroon chevrons, or both.**
LIGHT CONDITIONS: **Light shade**
SOIL/MOISTURE: **Moisture-retentive, fertile soil**

DESCRIPTION: Tovara is grown for its showy variegated leaves, which may reach 10" or more long. Since the leaves are easily damaged by wind, it is wise to site plants in a sheltered position. Tovara spreads readily by rhizomes and may need to be confined where space is limited. Striking as a ground cover in shaded spots under hemlocks or other dark evergreens. Interesting color echoes can be created by combining with zebra grass or variegated hosta.

Tradescantia virginiana
(spiderwort)

Tricyrtis latifolia
(Japanese toadlily)

Trillium discolor

RIGHT
Uvularia perfoliata
(strawbell)

Tradescantia virginiana

PRONUNCIATION: **trad-es-KAN-tee-a vir-jin-ee-AYE-na**

COMMON NAME: **Spiderwort**

HOMELAND: **Woods, meadows, and roadsides from Maine to Minnesota, south to Georgia, Tennessee, and Missouri**

HARDINESS: **USDA Zones 5–9**

SIZE: **18″–30″ tall; 12″–24″ across**

INTEREST: **Clusters of bright purple triangular flowers in late spring and summer; silky smooth, linear leaves.**

LIGHT CONDITIONS: **Full sun to very light shade**

SOIL/MOISTURE: **Well-drained soil of average or poor fertility**

DESCRIPTION: This species is seldom cultivated today, but has been used extensively in breeding programs, which have resulted in a wide range of popular named color forms. Though their habit is somewhat ungainly, they are useful in the garden for their ease of culture and long bloom season. Popular cultivars include white 'Osprey', lavender 'Blue Stone', mauve 'Pauline', and deep magenta 'Red Cloud'. Encourage blooming and avoid excessive vegetative growth by maintaining a lean soil. Each bloom opens for only one day.

Tricyrtis latifolia

PRONUNCIATION: **try-SER-tis lat-i-FO-lee-a**

COMMON NAME: **Japanese toadlily**

HOMELAND: **Japan**

HARDINESS: **USDA Zones 5–9**

SIZE: **15″–30″ tall; 18″–24″ across**

INTEREST: **In summer, branching clusters of yellow, purple-spotted flowers top wiry stems clothed with clasping, heart-shaped leaves.**

LIGHT CONDITIONS: **Partial shade to shade**

SOIL/MOISTURE: **Slightly acid, deep woodland soil**

DESCRIPTION: The erect stems of Japanese toadlily make good-sized clumps after several years. The interesting, unusual flowers demand a closer look and should be planted where they can be examined up close. Try to select a site out of the wind to avoid foliage damage. Combine with yellow variegated hostas for a good effect.

TRILLIUM DISCOLOR

PRONUNCIATION: TRILL-ee-um DIS-col-or

COMMON NAME: None

HOMELAND: Woods of North Carolina to Georgia

HARDINESS: USDA Zones 5–9

SIZE: 6"–12" tall; 12" across

INTEREST: In early spring, naked stems with a trio of speckled leaves, pale along the mid rib; a solitary yellow flower sits above.

LIGHT CONDITIONS: Light shade to shade

SOIL/MOISTURE: Woodsy, moisture-retentive soil

DESCRIPTION: Trillium are among the best known of our spring woodland flowers. The showiest and most easily recognizable is snow trillium, *Trillium grandiflorum;* others with good ornamental value include purple trillium, *T. erectum,* and nodding wake-robin, *T. catesbaei.* All have their leaf and flower parts in threes. Trillium is an endangered genus and may not be dug from the wild according to State and Federal laws; additionally, trilliums are legally protected from international commerce. Gardeners who buy plants from commercial sources should be certain that they have been nursery propagated not just nursery-grown or wild-collected.

UVULARIA PERFOLIATA

PRONUNCIATION: ewe-vew-LARE-ee-a per-fo-lee-AH-ta

COMMON NAME: Strawbell

HOMELAND: Rich, acid woods of Quebec to Ohio, Tennessee, Florida, and Louisiana

HARDINESS: USDA Zones 4–9

SIZE: 6"–12": tall; 12" across

INTEREST: Perfoliate, oval leaves to 3" long clothe wiry, branched stems; in spring, solitary, nodding, pale yellow 1" bells.

LIGHT CONDITIONS: Light shade to shade

SOIL/MOISTURE: Moisture-retentive, acid, humusy soil

DESCRIPTION: Strawbells are grown as much for their attractive, pleated leaves, arranged in a zigzig pattern, as for their rather shy flowers. They are interesting additions to early spring woodland gardens, where they make robust stands over time. Use the expanding foliage of lungworts and ferns to hide the shabby dieback of the strawbells.

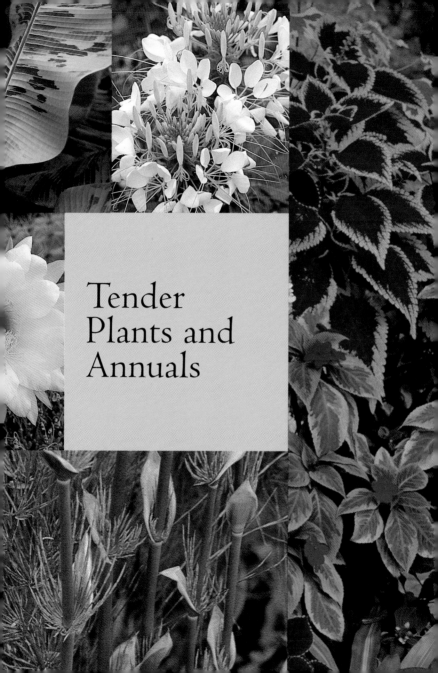

Tender
Plants and
Annuals

Tender Plants and Annuals

Annuals are plants that sprout from seed, grow, bloom, and set new seed all in one year—in cold climates usually from spring to fall, when they are cut to the ground by frost. Annuals are monocarpic, that is, they flower only once and die. Biennials, such as foxgloves and hollyhocks, are also monocarpic but take two years to bloom.

Very often, other plants represented as annuals in the nurseries are not monocarpic. They are herbaceous or woody perennials that originate in the warm regions of the world and therefore can not survive cold weather. Their forms are familiar: shrubs like tropical hibiscus and tender salvias; herbaceous plants like impatiens; and vines like passionflowers. If carried over outdoors, they will be struck dead as soon as temperatures dip below freezing. If they are given protection from the cold, however, they will live for years. Many can be lifted from the garden, cut back, and potted to spend the winter in sunny windows or greenhouses. Many house plants not only appreciate a summer outdoors, but can be incorporated into garden schemes for that season.

Because some other tender plants become dormant or semidormant over winter, they may be stored in cool, dark places, like a root cellar, a cool corner of the basement, or a spot in the garage where they are protected from freezing. Some tender shrubs can also be cut back and carried over in a semidormant state. Certain salvias, fuchsias, and geraniums (*Pelargonium* spp.) may be stored if kept dry and cool. Plants that grow from bulbs or bulblike tubers, corms, and rhizomes are also stored dry or nearly dry and cool.

Until recently, gardeners looked down upon old-fashioned plants, choosing instead the newest, brightest, biggest zinnia. What is most exciting today is that plants are changing as the demand expands. Cannas, for example, can be found with streaked foliage in yellow and green, rainbow shades, and burgundy striations. Coleus are appearing as named varieties and propagated vegetatively to retain their color characteristics. Tender perennials, annuals, and biennials afford more opportunities for us to explore every possibility of horticultural expression.

Alcea rosea
(hollyhock)

Aloe striata
(coral aloe)

Asparagus densiflorus
'Meyers' (foxtail fern)

Agave franzosinii
(blue agave)

AGAVE FRANZOSINII

PRONUNCIATION: **ag-AH-vay fran-zoz-IN-ee-aye**

COMMON NAME: **Blue agave**

HOMELAND: **Mexico**

HARDINESS: **USDA Zones 9–11**

SIZE: **4′–6′ tall and as wide**

INTEREST: **Dramatic, arching, steel blue fleshy leaves armed with triangular teeth along their edges; panicles of yellow flowers on 35′ stems.**

LIGHT CONDITIONS: **Full sun to very light shade**

SOIL/MOISTURE: **Very well-drained soil of average fertility**

DESCRIPTION: The imposing foliage of blue agave adds architectural impact to cactus and succulent gardens. In colder climates it is sometimes grown under glass. Individual leaves may exceed 1′ across at their widest point; the terminal spines are 2″–3″ long. Agave increases by offsets to develop into large colonies over time. Avoid placing where the hooked marginal spines could cause injury.

ALCEA ROSEA

PRONUNCIATION: **al-SEE-a ROW-see-a**

COMMON NAME: **Hollyhock**

HOMELAND: **China**

HARDINESS: **USDA Zones 2–9**

SIZE: **5′–10′ tall; 2′–3′ across**

INTEREST: **Clumps of coarse, hairy foliage; elegant stems, the top 2′ decorated with single or double 3″–5″ flowers in summer.**

LIGHT CONDITIONS: **Full sun**

SOIL/MOISTURE: **Moist, well-drained, fertile soil**

DESCRIPTION: The red-, pink-, yellow-, or maroon-flowered hollyhocks on the market today are probably of hybrid origin. There are several popular strains, such as the Indian Springs and Farmyard hybrids, both single flowered, and the double-flowered Chater's Double. The pastel-colored Majorette strain is also double, but only reaches 2′–3′ tall. Hollyhock is a favorite cottage garden plant, often combined with garden phlox, columbines, and honeysuckle. Susceptible to rust fungus; slugs and Japanese beetles may also need control. Protect from wind if possible.

ALOE STRIATA

PRONUNCIATION: AL-o stry-AH-ta
COMMON NAME: Coral aloe
HOMELAND: Southern Africa
HARDINESS: USDA Zones 9–11
SIZE: 2'–3' tall; 2' or more across

INTEREST: Rosettes of cupped, oval leaves on short reclining stems; widely branching clusters of coral-red flowers in late winter.
LIGHT CONDITIONS: Full sun to light shade
SOIL/MOISTURE: Very well-drained soil of average fertility

DESCRIPTION: The 18"–20" long leaves are fleshy and lack the marginal teeth common to many aloes. The individual 1" flowers are tubular. They bloom reliably each year and, where hardy, can be massed as a ground cover or planted with other succulents in rock gardens. In cooler climates they are grown in containers to decorate patios and conservatories.

ASPARAGUS DENSIFLORUS 'MEYERS'

PRONUNCIATION: as-PAR-ag-us denz-i-FLOR-us
COMMON NAME: Foxtail fern
HOMELAND: Of garden origin; species from southern Africa
HARDINESS: USDA Zones 9-10
SIZE: 18"–24" tall; 12"–24" across

INTEREST: Neat clumps of stiff, densely branched, plume-like stems to 2½" wide, and tapering to a point
LIGHT CONDITIONS: Light shade to shade
SOIL/MOISTURE: Moisture-retentive, well-drained, fertile soil

DESCRIPTION: Foxtail fern has long been grown as a Victorian parlor plant; it survives in containers large and small, in spite of less than ideal growing conditions. In climates mild enough for it to grow in the ground, it makes an attractive ground cover. The tuberous roots allow the plant to die down and become dormant during periods of drought or cold weather. Susceptible to mealy bugs and to red spider mites where humidity is low.

Callirhoe involucrata var. *tenuissima* (poppy mallow)

Canna × *generalis* 'Pretoria' (Bengal tiger canna)

Clarkia concinna ssp. *raichei* (small horned red ribbons)

LEFT
Coleus × *hybridus* [*Solenostemon scutellaroides*] 'Flirtin' Skirts' (fancy leaves)

Callirhoe involucrata var. *tenuissima*

PRONUNCIATION: **kall-ee-ROW-ee in-vol-OO-krata var. ten-ewe-iss-im-a**

COMMON NAME: **Poppy mallow, wine cup**

HOMELAND: **Mexico**

HARDINESS: **USDA Zones 6–9**

SIZE: **12″–18″ tall; 12″ or more across**

INTEREST: **Evergreen clumps of blue-green, finely dissected foliage; long-blooming, violet, cup-shaped flowers.**

LIGHT CONDITIONS: **Very light shade**

SOIL/MOISTURE: **Rich, well-drained soil**

DESCRIPTION: This variety of poppy mallow from the high altitudes of Mexico does not become invasive like the species. It is low to the ground and forms dense clumps which bloom all season long, in spite of heat and humidity.

Canna × *generalis* 'Pretoria'

PRONUNCIATION: **KANNA × jen-er-AL-is**

COMMON NAME: **Bengal tiger canna**

HOMELAND: **Of garden origin**

HARDINESS: **USDA Zones 7–11**

SIZE: **5″–6′ tall; 2′–3′ across**

INTEREST: **Dramatic, paddle-shaped, bright yellow foliage with veins picked out in green; summer spikes of showy orange flowers.**

LIGHT CONDITIONS: **Full sun**

SOIL/MOISTURE: **Moisture-retentive, deep, rich soil**

DESCRIPTION: Bengal tiger is one of the showiest of the canna cultivars on the market. The tropical-looking foliage is arresting, both as an accent plant or massed in the landscape. Excellent for containers, especially beside a pool or on a patio where a tropical effect is desired. Show off against a gray or dark green backdrop. Cannas are greedy feeders and should be well supplied with water and fertilizer. Be alert for slugs.

CLARKIA CONCINNA SSP. RAICHEI

PRONUNCIATION: **KLAR-kee-a kon-SINN-a ssp. RAY-chee-eye**

COMMON NAME: **Small horned red ribbons**

HOMELAND: **California**

HARDINESS: **USDA Zones 8–10**

SIZE: **1'-2' tall; 1' or more across**

INTEREST: **Long-blooming hot pink, four-petaled flowers**

LIGHT CONDITIONS: **Full sun to light shade**

SOIL/MOISTURE: **Well-drained, loose soil of average fertility**

DESCRIPTION: This subspecies of red ribbons was named for its discoverer Robert Raiche, horticulturist at UC Berkeley. It is an excellent annual plant for containers, or for any position where nonstop bloom is required. Its hot color is showy throughout the season.

COLEUS × HYBRIDUS 'FLIRTIN' SKIRTS'

PRONUNCIATION: **KO-lee-us × HIB-rid-us**

COMMON NAME: **Fancy leaves**

HOMELAND: **Of garden origin**

HARDINESS: **USDA Zones 10–11**

SIZE: **2'-3' tall; 1'-2' across**

INTEREST: **Rounded, burgundy-streaked green leaves edged with cream frills on sturdy, erect, branching plants**

LIGHT CONDITIONS: **Bright shade**

SOIL/MOISTURE: **Moisture-retentive, well-drained, fertile soil**

DESCRIPTION: The trend toward growing exotic-looking and sometimes even bizarre-looking foliage plants is complemented by the current popularity of container plantings. Coleus is a prime candidate for decorating containers of all kinds in regions where the plants would not survive the winter. The low-growing cultivars are also valuable for summer bedding, and in subtropical regions they make effective ground covers. Easy to grow and propagate, coleus can create a sparkling, tropical feel to a wide range of plant combinations. 'Flirtin' Skirts' is only one in a huge collection of recent introductions that illustrate well the common name, fancy leaves.

Echinocactus grusonii
(golden barrel cactus)

Gomphrena globosa (globe amaranth)

Impatiens sp. [*I. kew*]
(busy Lizzie)

RIGHT
Echium pininana
× *wildprettii*

ECHINOCACTUS GRUSONII

PRONUNCIATION: ee-KYE-no-kak-tus
  groo-SO-nee-aye

COMMON NAME: Golden barrel cactus

HOMELAND: Central Mexico

HARDINESS: USDA Zones 9–11

SIZE: 3′–4′ tall; 2′–2½′ across

INTEREST: Large, cylindrical cactus, deeply ridged and furnished with 3″ yellow spines along the outer ribs; yellow flowers in spring.

LIGHT CONDITIONS: Full sun to light shade

SOIL/MOISTURE: Very well-drained soil, amended with sand or crushed brick

DESCRIPTION: Golden barrel cactus grows slowly. It is best where protected from intense midday sun and needs regular watering in summer. Protect from sudden hard frosts in winter. The cup-shaped 1″–2″ flowers are borne in a circle at the top of the plant. This species is a popular subject for cactus and succulent gardens in favorable climates, and for containers where the climate is unfavorable.

ECHIUM PININANA × WILDPRETTII

PRONUNCIATION: EE-kee-um pin-in-AH-na
  × wild-PRET-tee-i

COMMON NAME: None

HOMELAND: Of garden origin

HARDINESS: USDA Zones 9–10

SIZE: 5′–6′ tall; 3′ or so across

INTEREST: A single, erect inflorescence, shaped like an Indian club; bright flowers interspersed with silvery bracts.

LIGHT CONDITIONS: Full sun

SOIL/MOISTURE: Well-drained, fertile, but not rich, soil

DESCRIPTION: This short-lived perennial hybrid usually dies after flowering. Though not as tall as some of the species, it is nevertheless imposing in bloom. The tapering spike is densely covered with flowers in reds, blues, and mauves. Plant to grow through shrubs and other sturdy plants for support.

GOMPHRENA GLOBOSA

PRONUNCIATION: **gom-FREE-na glo-BO-sa**
COMMON NAME: **Globe amaranth**
HOMELAND: **Central America; strains of garden origin**
HARDINESS: **Tender annual**
SIZE: **8″–30″ tall; 12″ or so across**

INTEREST: **Clover-like flowerheads in white, red, purple, and lavender, sometimes orange, over a very long bloom time**
LIGHT CONDITIONS: **Full sun**
SOIL/MOISTURE: **Well-drained, moisture-retentive, fertile soil**

DESCRIPTION: Globe amaranth is excellent massed, but also looks attractive grouped with perennials and other annuals. 'Strawberry Fields' ['Strawberry Fayre'] has scarlet flowers on 18″–24″ stems, and excels when planted in hot-colored designs, such as with calliopsis or 'Stella d'Oro' daylily. 'Lavender Lady' is a lovely soft lavender; accent it with *Verbena bonariensis* or lavender petunias. The 23″ Pomponette strain is available in separate colors. 'Dwarf Buddy' has magenta flowers on 8″ plants ('Buddy' tops out at 18″), and is suitable for edgings and containers. All cut and dry well. Water during dry spells to minimize mildew.

IMPATIENS SP. [*I. KEW*]

PRONUNCIATION: **im-PAT-shuns**
COMMON NAME: **Busy Lizzie, patience plant**
HOMELAND: **Of garden origin**
HARDINESS: **USDA Zones 10–11**
SIZE: **2′–3′ tall; 1′–3′ across**

INTEREST: **Short-spurred flowers in bright colors throughout the warm months; fleshy, variegated white, green, and pink leaves.**
LIGHT CONDITIONS: **Light shade to shade**
SOIL/MOISTURE: **Well-drained, moist soil**

DESCRIPTION: Impatiens are grown by the million as summer bedding plants in climates too cold for year-round growing. Most are hybrids of species such as the East African *I. wallerana,* New Guinea *I. schlechteri,* and *I. auicoma* from the Comoro Islands. Major seed companies offer lists of hybrid strains, selected for their habit, flowering period, or foliage. Many of the so-called New Guinea impatiens, largely derived from *I. schlechteri,* have brilliantly variegated foliage, and are grown as much for their leaves as for their flowers; they require sunnier conditions than the more familiar *I. wallerana* strains. Superior selections, either for foliage or blooms, must be propagated vegetatively.

Musa zebrina [*M. acuminata* 'Sumatrana'] (blood banana)

Phormium cookianum 'Tricolor' (New Zealand flax)

Salvia vanhouttii

LEFT
Lantana camara 'Miss Huff' (hardy lantana)

Lantana camara 'Miss Huff'

PRONUNCIATION: lan-TAN-a ca-MAR-a

COMMON NAME: Hardy lantana

HOMELAND: Garden selection; species from South America

HARDINESS: USDA Zones 7–10

SIZE: 2'–3' tall; 3'–10' across

INTEREST: In late spring until frost, an abundance of showy clusters of orange and pink flowers above rough, dark green oval leaves

LIGHT CONDITIONS: Full sun

SOIL/MOISTURE: Fertile, well-drained soil

DESCRIPTION: 'Miss Huff' is an almost sterile cultivar, which as a result continues to bloom without diverting energy to seed production. Established plants develop into large spreading clumps; cut back if they become rangy. Propagate by taking soft cuttings in summer. Other cultivars of lantana are hardy only in Zones 9 and warmer. They are often overwintered under glass in cold climates and make good subjects for sunrooms and conservatories. Plants trained as standards were popular "dot" plants in traditional bedding schemes.

Musa zebrina [M. acuminata 'Sumatrana']

PRONUNCIATION: MEW-za ze-BRY-na

COMMON NAME: Blood banana

HOMELAND: Tropical Southeast Asia

HARDINESS: USDA Zones 10–11

SIZE: 15'–20' tall; 6'–10' across

INTEREST: Huge, long-stalked, bluish-green leaves, irregularly blotched with blood red; "hands" of small green bananas ripen to yellow.

LIGHT CONDITIONS: Full sun to partial shade

SOIL/MOISTURE: Moist, fertile soil

DESCRIPTION: Blood bananas sucker freely to form small stands. In tropical regions they can be grown as ornamental accents or specimen plants outdoors; in cooler zones they must be grown in large containers and sheltered under glass during cold weather. Protect from wind to avoid shredding of the oar-shaped leaf blades. The "trunks" of banana plants are composed of ensheathing leaf bases, which never become woody, but are filled with columns of water; this allows for flexibility during periods of high winds in their native habitat. A good companion for palms and *Hibiscus rosa-sinensis.*

PHORMIUM COOKIANUM 'TRICOLOR'

PRONUNCIATION: FOR-mee-um cook-ee-AY-num

COMMON NAME: New Zealand flax

HOMELAND: Of garden origin; species from New Zealand

HARDINESS: USDA Zones 8–11

SIZE: 6'–12' tall; 3'–4' across

INTEREST: Evergreen clumps of sword-like, gray-green leaves, with reddish margins and edged with cream

LIGHT CONDITIONS: Sun to light shade

SOIL/MOISTURE: Well-drained, moist, fertile soil

DESCRIPTION: New Zealand flax are magnificent plants for introducing a vertical dimension into a garden. The stiff leaves remain handsome through much of the year, although they may sustain some winter damage. In late spring to summer, tall panicles of red, tubular flowers are held high above the rosette of foliage. Excellent for decorating terraces and patios in containers that can be protected in cold winter climates. Tolerant of seaside conditions, drought, and pollution. Propagate by division.

SALVIA VANHOUTTII

PRONUNCIATION: SAL-vee-a van-HOOT-ee-i

COMMON NAME: None

HOMELAND: Mexico and Central America

HARDINESS: USDA Zones 8–10

SIZE: 2'–3' tall and as wide

INTEREST: Spikes of tubular, carmine flowers with wine-colored bracts top this leafy perennial from late summer until frost.

LIGHT CONDITIONS: Light shade

SOIL/MOISTURE: Well-drained, moist, fertile soil

DESCRIPTION: This tender perennial makes a spectacular late-season display. In northern climates it is grown as an annual, or the roots are dug to overwinter under cover. A fine companion for ornamental grasses. In cold regions, set out established plants from quart-sized or larger pots as soon as the weather warms; delay may cause the plants to be cut down by frost before their floral display reaches its peak.

Climbing
Plants

Climbing Plants

Vines may be woody or herbaceous. We grow them for beautiful flowers or for foliage color and texture. But all vines have one thing in common—they grow up—climbing to fill a layer of the garden between shrubs and trees; covering a fence or trellis; or just functioning as another wonderful kind of collectible plant.

There are collectors who dote on clematis, seeking out every member of this wide and varied genus, those that bloom in early spring, and ones that flower in fall. Spring-blooming *Clematis montana* smells like vanilla; sweet autumn clematis (*C. terniflora* [*C. paniculata*]) like honey. These two grow tall—up to the third-story window. Familiar early-summer blooming hybrids have flowers up to ten inches across. The North American native, *C. virginiana* (virgin's bower) produces little fragrant stars about as big as a nickel.

There are shrubby clematis with leathery urn-shaped flowers that face the ground, and others with flaring blue bells.

In a place where you want to grow a vine, you do not necessarily have to grow the same one each year. There are plenty of annual and tender perennial vines to experience. *Ipomea lobata* (*Mina lobata*), relative of the common morning glory, is an easy-to-grow-from-seed climber with extraordinary flowers that open pale yellow and deepen to rich ruby red. Each flowering stalk has some of the different shades at the same time giving rise to its common name, Spanish flag. Cardinal creeper (*I.* × *multifida*) has fernlike leaves and tiny crimson trumpets.

Concurrently, there are perfectly hardy woody climbers that can become permanent additions to the garden. *Actinidia kolomikta* (hardy kiwi vine) has felty pointed bronze-green leaves that look as if they have been dipped halfway into white paint and then trimmed with pink. And did you know that there is a native North American wisteria, *Wisteria frutescens*?

There may not be as many collectors of vines as there are of other types of plants, but anyone with a small space should remember these climbers. You can pack a great variety of unusually varied flowering and foliage plants into a small garden when their habit is to grow up rather than out.

Actinidia kolomikta
(hardy kiwi)

Clematis 'Betty Corning'
(hybrid clematis)

Cobaea scandens
(cup and saucer vine)

RIGHT
Clematis 'Nelly Moser'
(hybrid clematis)

ACTINIDIA KOLOMIKTA

PRONUNCIATION: ak-tin-ID-ee-a ko-lo-MIK-ta

COMMON NAME: Hardy kiwi, Kolomikta actinidia

HOMELAND: Northeastern Asia, Japan, China

HARDINESS: USDA Zones 4–8

SIZE: 15'–30' tall and as wide, as allowed

INTEREST: Oval, 5″, dark green leaves with pink or white tips and clusters of fragrant white flowers in spring; greenish-yellow fruits in fall.

LIGHT CONDITIONS: Full sun to partial shade

SOIL/MOISTURE: Deep, rich, well-drained, fertile soil

DESCRIPTION: Like clematis, hardy kiwi climbs by twining around a support. For fruit production both male and female plants must be grown. The edible, oval, 1″ fruits are sweet-tasting and borne on old wood. Prune after flowering or in winter to keep within bounds. Leaf color is best in male plants, and under good light conditions but not intense heat. Seldom needs fertilizing.

CLEMATIS 'BETTY CORNING'

PRONUNCIATION: KLEM-a-tis

COMMON NAME: Hybrid clematis

HOMELAND: Of garden origin

HARDINESS: USDA Zones 3–9

SIZE: 8'–10' tall; 4'–6' wide

INTEREST: From early summer through fall, nodding, pale lilac flaring bells cover the plant; finely cut, deciduous foliage.

LIGHT CONDITIONS: Full sun to light shade

SOIL/MOISTURE: Moisture-retentive, rich, humusy soil

DESCRIPTION: 'Betty Corning' resulted from a cross of *C. crispa* × *C. viticella*. The 2½″–3″ flowers are lightly fragrant and bloom until late summer. The foliage remains attractive all season long. Prune in early spring, removing old wood. Position on a trellis, arbor, or other support facing south, east, or west for best flowering.

CLEMATIS 'NELLY MOSER'

PRONUNCIATION: KLEM-a-tis

COMMON NAME: Hybrid clematis

HOMELAND: Of garden origin

HARDINESS: USDA Zones 3–9

SIZE: 6'–8' tall; 4'–6' wide

INTEREST: Star-shaped 6"–8" mauve-pink flowers striped with deep pink in late spring and again in early fall

LIGHT CONDITIONS: Full sun to partial shade

SOIL/MOISTURE: Moisture-retentive, rich, humusy soil

DESCRIPTION: Clematis bloom best when their roots are kept cool and their tops are in sun. They are popular subjects for covering trellises, arbors, and fences, as well as camouflaging unsightly mailboxes and other utilitarian structures. 'Nelly Moser' is a vigorous grower and must be kept well watered and fed with a high potash fertilizer during the growing season; an organic mulch helps to retain moisture and keeps the roots cool. The flower color tends to fade in strong sun, so plant facing east or north, where there is some afternoon shade. Prune lightly after bloom time. There are numerous other large-flowered hybrid cultivars on the market in a wide range of colors.

COBAEA SCANDENS

PRONUNCIATION: KO-bea SCAN-dens

COMMON NAME: Cup and saucer vine, monastery bells

HOMELAND: Mexico and tropical America

HARDINESS: USDA Zones 9–11

SIZE: 20'–25' tall; 1'–2' across

INTEREST: In late summer and fall, solitary 2" bell-shaped purple or white flowers, with exserted, curved stamens

LIGHT CONDITIONS: Full sun to light shade

SOIL/MOISTURE: Well-drained, average soil

DESCRIPTION: This tender vine climbs by tendrils at the ends of the pinnate leaves. In northern climates it is grown outdoors as an annual, but it also makes a good sunroom or conservatory plant. Excellent for covering arches, trellises, or arbors, or for growing through or on shrubs and evergreens.

Passiflora alata
'Ruby Glow'
(passionflower)

Vitis vinifer 'Purpurea'
(claret vine)

Wisteria floribunda
(Japanese wisteria)

LEFT
Hedera helix 'Buttercup'
(Buttercup English ivy)

HEDERA HELIX 'BUTTERCUP'

PRONUNCIATION: HED-er-a HE-liks

COMMON NAME: Buttercup English ivy

HOMELAND: Of garden origin; species from the Caucasus Mountains

HARDINESS: USDA Zones 4–9

SIZE: 70′–90′ or more

INTEREST: Evergreen, lobed leaves, often accented with white or yellowish veins. Unlobed adult foliage and black berries on mature plants only.

LIGHT CONDITIONS: Partial shade to shade

SOIL/MOISTURE: Well-drained, moist, acid or alkaline organic soil

DESCRIPTION: There are countless forms, selections, and cultivars of this popular vine. 'Buttercup' has clear yellow new growth deepening to grass green by summer's end. The next year, the leaves are deep green, but fresh foliage emerges to light up a shady spot. Some of the other selections include hardy forms: '238th Street', which does not succumb to winter burn; 'Baltica', which has small leaves; and 'Bulgaria', possibly the hardiest. 'Glacier' is a popular cultivar with gray-green foliage variegated with white. The green leaves of 'Goldheart' have a yellow blotch in the center of each. English ivy is effective when climbing and may also be used as a ground cover, but prune to keep within bounds.

PASSIFLORA ALATA 'RUBY GLOW'

PRONUNCIATION: pass-i-FLOR-a

COMMON NAME: Passionflower

HOMELAND: Of garden origin; species from Brazil and northeastern Peru

HARDINESS: USDA Zones 10–11

SIZE: 20′–30′ vine

INTEREST: Winged stems with oval 3″–6″ leaves; fragrant 6″ crimson flowers with a purple corona; yellow 4″–6″ oval fruits.

LIGHT CONDITIONS: Sun to partial shade

SOIL/MOISTURE: Well-drained soil of average fertility

DESCRIPTION: Passionflowers climb by means of twining tendrils. They make handsome climbers for covering walls and trellises in frost-free climates, but elsewhere they make excellent container subjects for conservatories and sunrooms. Shade from intense sun to protect from leaf burn. Avoid high-nitrogen fertilizers; they promote excessive vegetative growth at the expense of flowers and fruit. Prune out weak growth in spring and again after flowering, if necessary. This species and its cultivars are among the best for fruit production.

VITIS VINIFER 'PURPUREA'

PRONUNCIATION: VI-tis vin-IF-er-a

COMMON NAME: Claret vine, Teinturier grape, Purple-leaved grape

HOMELAND: Of garden origin; species from Caucasus Mountains

HARDINESS: USDA Zones 5–9

SIZE: To 30' tall; 6'–15' across

INTEREST: Deeply lobed, rounded 6" leaves, claret red when young, turning deep purple-red in fall

LIGHT CONDITIONS: Full sun to very light shade

SOIL/MOISTURE: Well-drained soil of average fertility

DESCRIPTION: This attractive deciduous vine clothes a wall or trellis rapidly once established. The young foliage, brilliantly colored in spring, becomes dull in summer's heat. Before the leaves drop in fall they again take on spectacular color. The bunches of blue-black fruit are covered with a whiteish bloom. The best foliage color occurs in full sun. An especially good accent for a gray stone wall.

WISTERIA FLORIBUNDA

PRONUNCIATION: wis-TEE-ree-a flor-i-BUN-da

COMMON NAME: Japanese wisteria

HOMELAND: Japan

HARDINESS: USDA Zones 4–9

SIZE: 25'–30' or more tall; 4'–10' across or more

INTEREST: A fast-growing vine bearing violet-blue pea flowers in slender racemes to 20" long in mid-spring

LIGHT CONDITIONS: Full sun

SOIL/MOISTURE: Deep, well-drained, moist, neutral soil

DESCRIPTION: Similar to Chinese wisteria, which twines counterclockwise, Japanese wisteria climbs by stems that twine clockwise, eventually developing into large, twisted woody trunks. These trunks are very vigorous and can tear down trellises, fences, and arbors. Considerable thought should be given to siting this vine. The lightly fragrant flowers bloom before or just at the time when the foliage emerges and open from the base of the raceme towards the tip. Spur-prune hard to 3–4 buds in late winter. Japanese wisteria is sometimes challenging to get to bloom; root confinement in a container and/or severe root pruning may be required. Avoid nitrogen fertilizers; they encourage excessive vegetative growth.

Appendix

Mail-Order
Nurseries

Adamgrove Nursery
Route 1, Box 246
California, MO 65018
Hemerocallis, *irises,*
peonies. Catalog: $3

Alpen Gardens
173 Lawrence Lane
Kalispell, MT
59901-4633
(406) 257-2540
Dahlia tubers

Jacques Amand, Bulb
  Specialists
PO Box 59001
Potomac, MD 20859
(800) 452-5414
Spring and summer blooming
bulbs

Ambergate Gardens
8730 County Rd. 43
Chaska, MN 55318
(612) 443-2248
Hostas, unusual perennials,
Martagon lilies. Catalog: $2

Anderson Iris Gardens
22179 Keather Ave. N.
Forest Lake, MN 55025
(612) 433-5268
Bearded iris, peonies.
Catalog: $1

Antique Rose Emporium
9300 Lueckemeyer
Brenham, TX 77833
(409) 836-9051
Old garden roses. Catalog: $5

Antonelli Brothers, Inc.
2545 Capitola Rd.
Santa Cruz, CA 95062
(408) 475-5222
Tuberous begonias, fuchsias.
Catalog: $1

Appalachian Gardens
PO Box 82
Waynesboro, PA 17268
(717) 762-4312
Conifers, flowering
shrubs, trees.

Arborvillage Farm
  Nursery
PO Box 227
Holt, MO 64048
(816) 264-3911
Shrubs, trees. Catalog: $1

Arrowhead Alpines
PO Box 857
Flowerville, MI 48836
(517) 223-3581
Conifers, wildflowers, ferns,
alpines. Catalog: $2

Arrowhead Nursery
5030 Watia Rd.
Bryson City, NC
28713-9683
Trees, shrubs native to
Southeast. List

B & D Lilies
330 P St.
Port Townsend, WA
98368
(360) 385-1738 *Lilies.*
Catalog: $3

Bio-Quest International
1781 Glen Oak Dr.
Santa Barbara, CA
93108
(805) 969-4072
Rare Clivia, *amaryllis
hybrids*

Bluestone Perennials
7211 Middle Ridge Rd.
Madison, OH 44057
(800) 852-5243
Herbaceous perennials, shrubs

Borbeleta Gardens
15980 Canby Ave.
Fairbault, MN 55021
(507) 334-2807
Iris, Hemerocallis,
Asiatic lilies. Catalog: $3

The Bovees Nursery
1737 SW Coronado
Portland, OR 97219
(503) 244-9341
*Flowering shrubs, trees,
perennials, vines, tropical
rhododendrons. Catalog: $2*

Bonnie Brae Gardens
1105 SE Christensen Rd.
Corbett, OR 97019
(503) 695-5190
*Daffodils. List: send long
SASE*

Brand Peony Farm
PO Box 842
Saint Cloud, MN
56302
*Peonies, especially heirloom
varieties. Catalog: $1*

Briarwood Gardens
14 Gully Lane
East Sandwich, MA
02537
Rhododendrons. Catalog: $1

Brown's Kalmia and
  Azalea Nursery
8527 Semiahmoo Dr.
Blaine, WA 98230
(360) 371-5551
Kalmia, *azaleas. List: $1*

Busse Gardens
13579 Tenth St., NW
Cokato, MN
55321-9426
(320) 286-2654
Herbaceous perennials,
Hemerocallis, *hostas, iris,
peonies. Catalog: $2*

Cactus by Dodie
934 Mettler Rd.
Lodi, CA 95242
(209) 368-3692
*Cacti and succulents.
Catalog: $2*

California Carnivores
7020 Trenton-
Healdsburg Rd.
Forestville, CA 95436
(707) 838-1630
*Carnivorous plants.
Catalog: $2*

Camellia Forest Nursery
125 Carolina Forest Rd.
Chapel Hill, NC 27516
*Conifers, flowering shrubs,
trees, some herbaceous plants,
camellias. Catalog: $2*

Campanula Connoisseur
702 Traver Trail
Glenwood Springs, CO
81601
Campanula. *Catalog: $1*

Caprice Nursery
15425 Southwest
Pleasant Hill Rd.
Sherwood, OR 97140
(503) 625-7241
*Peonies, Japanese and
Siberian irises,*
Hemerocallis, Hostas.
Catalog: $2

Carroll Gardens
PO Box 310
Westminster, MD
21158
(410) 848-5422
*Flowering shrubs, trees,
herbaceous perennials, grasses,
hardy ferns. Catalog: $3*

Cascade Daffodils
PO Box 10626
White Bear Lake, MN
55110-0626
(612) 426-9616
*Collector's miniature and
standard daffodils.
Catalog: $2*

Cascade Forestry
  Nursery
22033 Fillmore Rd.
Cascade, IA 52033
(319) 852-3042
Conifers, shrubs, and trees

Christa's Cactus
529 West Pima
Coolidge, AZ 85228
(520) 723-4185
Desert trees, shrubs, succu-
lents, cacti. Catalog: $1

Collector's Nursery
16804 NE 102d Ave.
Battle Ground, WA
98604
(360) 574-3832
Unusual conifers, flowering
shrubs, trees, herbaceous peren-
nials, vines, alpines, dwarf
conifers, Gentiana,
Tricyrtis, species Iris.
Catalog: $2

Colorado Alpines, Inc.
PO Box 2708
Avon, CO 81620
(970) 949-6464
Dwarf conifers, alpines, native
shrubs, trees, plants of the west

Companion Plants
7247 North Coolville
  Ridge Rd.
Athens, OH 45701
(614) 592-4643
Woodland plants, perennials,
herbs, scented geraniums.
Catalog: $3

The Compleat Garden
  Clematis Nursery
217 Argilla Rd.
Ipswich, MA
 01938-2617
(508) 356-3197
Clematis. Catalog: $3

Cooley's Gardens
PO Box 126
Silverton, OR 97381
(503) 873-5463
Bearded iris. Catalog: $5

Cooper's Garden
2345 Decatur Ave.
N. Golden Valley, MN
55427
(612) 591-0495
Herbaceous perennials, Iris.
Catalog: $1

Country Bloomers
  Nursery
RR 2
Udall, KS 67146
(316) 986-5518
Old garden roses, miniatures,
some modern roses. List: send
long SASE

Country Cottage
Route 2, Box 130
Sedgwick, KS 67135
Groundcover succulents. List:
send long SASE

Cricket Hill Garden
670 Walnut Hill Rd.
Thomaston, CT 06787
(860) 283-1042
Chinese tree peonies.
Catalog: $2

Crownsville Nursery
PO Box 797
Crownsville, MD 21032
(410) 849-3143
Woody plants, herbaceous
perennials, grasses, ferns,
Hemerocallis, hostas.
Catalog: $2

The Cummins Garden
22 Robertsville Rd.
Marlboro, NJ 07746
(732) 536-2591
Dwarf conifers,
Rhododendron, *azalea,*
Pieris, Kalmia, *heathers.*
Catalog: $2

Cycad Gardens
4524 Toland Way
Los Angeles, CA 90041
(213) 255-6651
Cycads. List: send long SASE

Daffodil Mart
Route 3, Box 794
Gloucester, VA 23061
(804) 693-6339
Vast list of Narcissus *vari-*
eties; also tulips, crocuses, alli-
ums, bulbs. Catalog: $1

Desert Nursery
1301 S. Copper St.
Deming, NM 88030
(505) 546-6264
Succulents and hardy cacti.
List: send long SASE

Desert Theatre
17 Behler Rd.
Watsonville, CA 95076
(408) 728-5513
South American and African
succulents and cacti.
Catalog: $2

Dooley Gardens
212 North High Dr.
Hutchinson, MN 55350
(320) 587-3050
Chrysanthemums

Jim Duggan Flower
  Nursery
1452 Santa Fe Dr.
Encinitas, CA 92024
(619) 943-1658
South African bulbs.
Catalog: $2

Eastern Plant Specialties
PO Box 226
Georgetown, ME
04548
(207) 371-2888
Dwarf conifers, flowering
shrubs, trees, Kalmia,
Rhododendron, *azalea.*
Catalog: $2

Fairweather Gardens
PO Box 330
Sheppards Mill Rd.
Greenwich, NJ 08323
(609) 451-6261
Woody plants, flowering
shrubs and trees

Fancy Fronds
PO Box 1090
Gold Bar, WA 98251
Hardy and temperate ferns,
many new introductions.
Catalog: $2

Field House Alpines
6730 W. Mercer Way
Mercer Island, WA
98040
Alpine and rock garden
seeds.Catalog: $2

Fieldstone Gardens
620 Quaker Lane
Vassalboro, ME
04989-9713
(207) 923-3836
Herbaceous perennials, herbs,
alpines, groundcovers.
Catalog: $2

Foliage Gardens
2003 128th Avenue SE
Bellevue, WA 98005
(206) 747-2998
Ferns and dwarf Japanese
maple cultivars. Catalog: $2

Forest Farm
990 Tetherow Rd.
Williams, OR
97544-9599
Conifers, flowering shrubs,
trees, herbaceous plants.
Catalog: $3

Fox Hill Farm
434 W. Michigan Ave.
Parma, MI
49269-0009
(517) 531-3179
Herbs, scented geraniums

The Fragrant Path
PO Box 328
Fort Calhoun, NE
68023
Fragrant perennials, annu-
als, herbs, vines (rare and
heirloom). Catalog: $2

Garden Place
PO Box 388
Mentor, OH
44061-0388
(216) 255-3705
Groundcovers, perennials,
grasses. Catalog: $1

Gilson Gardens
3059 U.S. Route 20
PO Box 277
Perry, OH 44081
(216) 259-5252
Low-growing shrubs, peren-
nials, vines, and sedums

Girard Nurseries
PO Box 428
Geneva, OH 44041
(216) 466-2881
Conifers, flowering shrubs,
trees, groundcovers, perenni-
als, vines

Glasshouse Works
  Greenhouses
Church St.
PO Box 97
Stewart, OH
45778-0097
(614) 662-2142
Tropical and subtropical
plants, ferns, succulents,
shrubs, trees, dwarf conifers,
tender and hardy perennials,
variegated plants.
Catalog: $2

Goodwin Creek Gardens
PO Box 83
Williams, OR 97544
(541) 846-7357
Herbs, fragrant plants, plants for hummingbirds and butterflies. Catalog: $1

Gossler Farms Nursery
1200 Weaver Rd.
Springfield, OR
97478-9691
(541) 746-3922
Conifers, flowering shrubs, trees, Hamamelis, Magnolia, Rhododendron. Catalog: $2

The Gourd Garden
4808 E. Country Rd.
30-A
Santa Rosa Beach, FL
32459
(904) 231-2007
Gourd and herb seed. List: send long SASE

The Green Escape
PO Box 1417
Palm Harbor, FL 34682
(813) 784-1991
Rare and uncommon palms for the conservatory and cold hardy. Catalog: $6

GreenLady Gardens
aka Skittone Bulbs
1415 Eucalyptus
San Francisco, CA
94132
(415) 753-3332
Wide variety of species and bulbs. Catalog: $3

Greenmantle Nursery
3010 Ettersburg Rd.
Garberville, CA 95440
(707) 986-7504
Old Garden roses. List: send long SASE

Greer Gardens
1280 Goodpasture
Island Rd.
Eugene, OR
97401-1794
Conifers, flowering shrubs, trees, perennials, grasses, ferns, bonsai plants, azalea, Rhododendron. Catalog: $3

Grigsby Cactus Gardens
2326-2354 Bella Vista
Dr.
Vista, CA 92084-7836
(760) 727-1323
Rare succulents and cacti. Catalog: $2

Heard Gardens, Ltd.
5355 Merle Hay Rd.
Johnston, IA 50131
(515) 276-4533
Lilacs. Catalog: $2

Heaths and Heathers
E. 502 Haskell Hill Rd.
Shelton, WA
98584-8429
(360) 427-5318
Heaths and heathers. List: send long SASE

Heirloom Garden Seeds
PO Box 138
Guerneville, CA 95446
Herbs and heirloom flowers. Catalog: $2.50

Heirloom Old Garden
  Roses
24062 NE Riverside Dr.
St. Paul, OR 97137
(503) 538-1576
Old garden and English roses. Catalog: $5

Heritage Rose Gardens
16831 Mitchell Creek Dr.
Fort Bragg, CA
95437-8727
(707) 964-3748
Old garden roses. Catalog: $1.50

Heronswood Nursery
7530 288th St. NE
Kingston, WA 98346
(206) 297-4172
Conifers, flowering shrubs, trees, herbaceous plants. Catalog: $3

High Country Rosarium
1717 Downing St.
Denver, CO 80209
(303) 832-4026
Old garden roses, roses for a variety of conditions. Catalog: $1

Hildenbrandt's Iris
   Gardens
1710 Cleveland St.
Lexington, NE
68850-2721
(308) 324-4334
*Hostas, peonies, bearded iris,
poppies. List: send long SASE*

Holly Haven Hybrids
136 Sanwood Rd.
Knoxville, TN
37923-5564
*Hollies. List: send long
SASE*

Holly Lane Iris
   Gardens
10930 Holly Lane
Osseo, MN 55369
(612) 420-4876
*Bearded iris, Siberian iris,
peonies, Hemerocallis,
Hosta*

Hollyvale Farm
PO Box 434
Hoquiam, WA 98520
Hollies. Catalog: $5

J. L. Hudson,
Seedsman
PO Box 1058
Redwood City, CA
94064
*No phone. No visits. Seeds
only—but seeds of just
about everything from all
over the world. Catalog: $1*

Huff's Garden Mums
PO Box 187
Burlington, KS 66839-
0187
(800) 279-4675
Chrysanthemums

Intermountain Cactus
2344 S. Redwood Rd.
Salt Lake City, UT
84119
(801) 966-7176
*Hardy cactus. List: send long
SASE*

Iris Country
6219 Topaz St. NE
Brooks, OR 97305
(503) 393-4739, 6:00
A.M. or eves., PST
*Iris: beardless, species, his-
toric bearded, modern ultra-
hardy bearded.
Catalog: $1.50*

Iris Test Gardens
James and Janet Leifer
1102 Endicott-
   St. John Rd.
St. John, WA 99171
(509) 648-3873
*Unusual bearded iris.
Catalog: $1*

Ivies of the World
PO Box 408
Weirsdale, FL
32195-0408
(352) 821-2201
Ivy. Catalog: $2

Joy Creek Nursery
20300 NW Watson Rd.
Scappoose, OR 97056
(503) 543-7474
*Shrubs, herbaceous perenni-
als, alpines, grasses.
Catalog: $2*

Kartuz Greenhouses
1408 Sunset Dr.
Vista, CA 92083-6531
(619) 941-3613
*Tropical and subtropical
plants, such as begonias,
Gesneriad, Passiflora.
Catalog: $2*

Kelleygreen Rhodo-
   dendron Nursery
185 Roaring Camp Ln.
Drain, OR 97435
(541) 836-2290
Rhododendron, *Japanese
maples, azalea,* Pieris, *and*
Kalmia. *Catalog: $1.25*

Kelly's Plant World
10266 E. Princeton
Sanger, CA 93657
(209) 294-7676
*Rare and unusual plants,
summer-blooming bulbs,*
Canna, Lycoris,
Crinum, *also trees and
shrubs. Catalog: $1*

Klehm Nursery
4210 N. Duncan Rd.
Champagne, IL 61821
(800) 553-3715
*Herbaceous perennials,
ferns, Siberian iris,*
Hemerocallis, *hostas,
peonies. Catalog: $4*

Lamb Nurseries
Route 1, Box 460B
Longbeach, WA 98631
(360) 642-4856
*Alpines and perennials, also
groundcovers, succulents, vines,
and flowering shrubs.
Catalog: $2*

Lamtree Farm
2323 Copeland Rd.
Warrensville, NC 28693
(910) 385-6144
*Native propagated trees and
shrubs:* Franklinia,
Stewartia, Styrax,
Halesia, Rhododendron,
azalea, Kalmia. *Catalog: $2*

Landscape Alternatives
1705 St. Alban's St.
Rooksville, MN 55113
(612) 488-3142
*Native U.S. wildflowers.
Growing Guide: $2*

Las Pilitas Nursery
3232 Las Pilitas Rd.
Santa Margarita, CA
93453
(805) 438-5992
*California native plants,
plants for special conditions.
Catalog: $8 (price list is free)*

Lilypons Water Gardens
6800 Lilypons Rd.
PO Box 10
Buckeystown, MD
21717-0010
(301) 874-5133
Aquatic plants. Catalog: $5

Logee's Greenhouses
141 North St.
Danielson, CT 06239
(860) 774-8038
*Tropical and subtropical
shrubs, vines, tender perennials, begonias, geraniums*

Louisiana Nursery
5853 Highway 182
Opelousas, LA 70570
(318) 948-3696
*Catalogs:
$6 Magnolias, perennials, and
woody plants
$4* Iris *and* Hemerocallis
*$5 Crinum and other rare
bulbs
$4 Fruiting trees, shrubs, and
vines
$4* Hydrangea
*$4 Bamboos and ornamental
grasses
$3* Clivia *list
$29.50 for all*

Lowe's Own-Root
Roses
6 Sheffield Rd.
Nashua, NH
03062-3028
(603) 888-2214
*Old garden roses, shrubs,
climbers, ramblers, and custom
grafted. Catalog: $3*

Mad River Imports
PO Box 1685
Fayston, VT 05660
(802) 496-3004
*Spring and summer blooming
bulbs*

Maple Tree Gardens
PO Box 547
Ponca, NE 68770-0547
(402) 755-2615
Maple trees, bearded iris,
Hemerocallis, Hosta.
Catalog: $1

Maryland Aquatic
Nurseries
3427 N. Furnace Rd.
Jarrettsville, MD 21084
(410) 557-7615
*Aquatic and waterside plants.
Catalog: $2*

Mary's Plant Farm
2410 Lanes Mill Rd.
Hamilton, OH 45013
(513) 892-2055
*Flowering shrubs, perennials,
ferns, iris, grasses, native
plants. Catalog: $1*

Mendocino Heirloom
Roses
PO Box 670
Mendocino, CA 95460
(707) 937-0963
*Antique, species, and unusual
roses. Catalog: $1*

Midwest Cactus
PO Box 163
New Melle, MO 63365
(314) 828-5389
*Hardy cacti, sedums, and yuc-
cas. Catalog: $2*

Mileager's Gardens
4838 Douglas Ave.
Racine, WI
53402-2498
(414) 639-2371
Roses, herbaceous perennials, vines, grasses. Catalog: $1

Miniature Plant
  Kingdom
4125 Harrison Grade
  Rd.
Sebastopol, CA 95472
(707) 874-2233
Dwarf conifers, bonsai suitable trees and shrubs, miniature roses, some perennials and alpines. Catalog: $2.50

Mountain Maples
PO Box 1329
54561 Registrar's
  Guest Rd.
Laytonville, CA
95454-1329
(707) 984-6522
Japanese and other maples. Catalog: $2

Mount Tahoma
  Nursery
28111 112th Ave.
E. Graham, WA 98338
(206) 847-9827
Small shrubs, and primarily alpines. Catalog: $1

Nature's Curiosity
  Shop
3551 Evening Canyon
  Rd.
Oceanside, CA 92056
Variegated plants, succulents. Catalog: $1

Neon Palm Nursery
3525 Stony Point Rd.
Santa Rosa, CA 95407
(707) 585-8100
Subtropical palms and cycads. Catalog: $1

New Peony Farm
PO Box 18235
St. Paul, MN 55118
(612) 457-8994
Peonies

Niche Gardens
1111 Dawson Rd.
Chapel Hill, NC
27516
(919) 967-0078
Natives of the southeastern U.S. Catalog: $3

Nichol's Garden
  Nursery, Inc.
1190 N. Pacific Hwy.
Albany, OR 97321
(541) 928-9280
Herbs, mints

Nurseries at North
  Glen
Route 2 Box 2700
Glen Saint Mary, FL
32040
(904) 259-2754
Hardy palms and cycads

Oakes Daylillies
8204 Monday Rd.
Corryton, TN 37721
(423) 687-3770
Hemerocallis

Oikos Tree Crops
PO Box 19425
Kalamazoo, MI
49019-0425
(616) 624-6233
Fruit and nut trees, oaks. Catalog: $1

Old House Gardens
536 Third St.
Ann Arbor, MI
48103-4957
(313) 995-1486
Heirloom bulbs. Catalog: $1

Orgel's Orchids
18950 Southwest
136th St.
Miami, FL
33196-1942
(305) 233-7168
Carnivorous plants. List: send long SASE

Peter Pauls Nurseries
4665 Chapin Rd.
Canandaigua, NY
14424-8713
(716) 394-7397
Carnivorous plants

Perennial Pleasures
  Nursery
2 Brick House Rd.
East Hardwick, VT
05836
(802) 472-5104
Perennials and herbs for historic restoration plantings. Catalog: $2

Piccadilly Farm 1971
Whippoorwill Rd.
Bishop, GA 30621
(706) 769-6516
Hosta, Helleborus
Catalog: $1

Plant Delights Nursery
9241 Sauls Rd.
Raleigh, NC 27603
(919) 772-4794
Conifers, flowering shrubs,
herbaceous perennials, grasses,
hostas. Catalog: $2

Plants of the Southwest
Agua Fria
Route 6, Box 11A
Santa Fe, NM 87505
(505) 438-8888
Flowering shrubs, trees, herba-
ceous perennials, vegetables,
herbs, xerophytes, penstemons.
Catalog: $3.50

Plants of the Wild
PO Box 866
Tekoa, WA 99033
(509) 284-2848
Natives of the Pacific
Northwest. Catalog: $1

Pond Doctor
HC 65, Box 265
Kingston, AR 72742
(501) 665-2232
Aquatic plants. Catalog: $2

Prairie Moon Nursery
Route 3, Box 163
Winona, MN 55987
(507) 452-5231
Native plants of midwest U.S.
Catalog: $2

Prairie Nursery
PO Box 306
Westfield, WI 53964
(608) 296-3679
Native perennials and grasses
of the U.S. prairie

Primrose Path
R.D. 2, Box 110
Scottdale, PA 15683
(412) 887-6756
Herbaceous perennials, alpines.
Catalog: $2

Quality Plants
6792 Buell Rd.
Igo, CA 96047
(916) 467-3426
Succulents and Lewesia
species. Catalog

Rare Conifer Nursery
PO Box 100
Potter Valley, CA
95469
Conifers

Rare Plant Research
9527 Southeast Wichita
Milwaukee, OR 97222
Rare succulents. List: send
long SASE

Rarifolia
Kintnersville, PA 18930
(215) 847-8208
Dwarf conifers, Japanese
maples. Catalog: $3

Reath's Nursery
County Rd. 577N-195
Vulcan, MI 49892
(906) 563-9777
Peonies. Catalog: $2

Riverdale Iris Gardens
PO Box 524
Rockford, MN 55373
(612) 477-4859
Siberian, hardy dwarf, tall,
and bearded iris. Catalog: $1

Robinett Bulb Farm
PO Box 1306
Sebastopol, CA
95473-1306
West coast native bulbs:
Allium, Brodiaea,
Calochortus, *others*
List: send long SASE

Rocknoll Nursery
7812 Mad River Rd.
Hillsboro, OH 45133
(614) 454-3018
Alpines, rock, and U.S.
natives. Catalog: $1

Rock Spray Nursery
PO Box 693
Truro, MA 02666
(508) 349-6769
Heaths and heathers.
Catalog: $2

Rocky Mountain Rare
  Plants
PO Box 200483
Denver, CO
80220-0483
Seeds only, alpines.
Catalog: $1

Roses of Yesterday and
Today
802 Brown's Valley Rd.
Watsonville, CA
95076-0398
(408) 724-3537
*Old garden roses, hybrid
teas. Catalog: $4*

Roslyn Nursery
211 Burrs Lane
Dix Hills, NY 11746
(516) 643-9347
*Conifers, flowering shrubs,
trees, herbaceous perennials,
hostas, ferns, azalea,*
Rhododendron,
Kalmia, Pieris,
Viburnum, Camellia,
Ilex. *Catalog: $3*

Russell Graham
4030 Eagle Crest Rd.
NW
Salem, OR 97304
(503) 362-1135
*Shrubs, perennials, ferns.
Catalog: $2*

Sandy Mush Herb
Nursery
316 Surrett Cove Rd.
Leicester, NC
28748-5517
(704) 683-2014
*Flowering shrubs, trees,
perennials, and herbs.
Catalog: $5*

Savory's Gardens, Inc.
5300 Whiting Ave.
Edina, MN 55439
(612) 941-8755
Perennials, shade plants,
Hemerocallis, Hosta.
Catalog: $2

John Scheepers, Inc.
PO Box 700
Bantam, CT 06750
(860) 567-0838
*Large selection of bulbs,
standard varieties. Catalog:
free with $25 minimum
order*

Schreiner's Gardens
3625 Quinaby Rd. NE
Salem, OR 97303
(503) 393-3232
Iris. *Catalog: $5*

Seeds of Change
1364 Rufina Circle #5
Santa Fe, NM 87501
*Vegetable and annuals, heir-
loom varieties. Catalog: $3*

Select Seeds
180 Stickney Rd.
Union, CT
06076-4617
(860) 684-9310
*Old-fashioned flowers and
seeds. Catalog: $1*

Sevald Nursery
4937 Third Ave. S.
Minneapolis, MN
55409
(612) 822-3279
*Herbaceous peonies.
Catalog: $2*

Shady Oaks Nursery
112 Tenth Ave. SE
Waseca, MN 56093
(507) 835-5033
*Shrubs, herbaceous perenni-
als, hostas, native plants,
ferns, groundcovers for shade.
Catalog: $2.50*

Shein's Cactus
3360 Drew St.
Marina, CA 93933
(408) 384-7765
*Cactus and succulents.
Catalog: $1*

Shooting Star Nursery
444 Bates Rd.
Frankfort, KY 40601
(502) 233-1679
*Native shrubs, trees, herba-
ceous perennials, grasses*

Siskiyou Rare Plant
Nursery
2825 Cummings Rd.
Medford, OR 97501
(541) 772-6846
*Dwarf conifers; dwarf
shrubs and trees; alpine rock,
woodland plants; hardy
ferns. Catalog: $3*

Stallings Nursery
910 Encinitas Blvd.
Encinitas, CA 92024
(619) 753-3079
*Tropical and subtropical
plants. Catalog: $3*

Story House Herb Farm
Route 7, Box 246
Murray, KY 42071
(502) 753-4158
Herbs. Catalog: $2

Sunlight Gardens, Inc.
174 Golden Lane
Andersonville, TN
37705
(423) 494-8237
Trees, flowering shrubs, herbaceous perennials, ferns— native to eastern North America. Catalog: $3

Sunnybrook Farms
  Nursery
9448 Mayfield Rd.
Chesterland, OH 44026
(216) 729-7232
Herbs, perennials, ivies, and hostas. Catalog: $2

Sunnyslope Gardens
8638 Huntington Dr.
San Gabriel, CA 91775
(818) 287-4071
Chrysanthemums

Sunshine Farm and
  Gardens
Route 5
Renick, WV 24966
(304) 497-3163
Very rare and exceptional plants. List: send long SASE

Swan Island Dahlias
PO Box 700
Canby, OR 97013
(503) 266-7711
Dahlias. Catalog: $3

Thompson and Morgan
PO Box 1308
Jackson, NJ
08527-0308
(908) 363-2225
Mostly seeds, some plants

Tranquil Lake Nursery
45 River St.
Rehobeth, MA
02769-1359
Hemerocallis, *Japanese and Siberian iris. Catalog: $1*

Transplant Nursery
1586 Parkertown Rd.
Lavonia, GA 30553
(706) 356-8947
Southeastern U.S. natives, Rhododendron, Camellia, *chiefly* Azalea. *Catalog: $1*

William Tricker, Inc.
7125 Tanglewood Dr.
Independence, OH
44131
(216) 524-3491
Aquatic plants, dwarf water lilies. Catalog: $2

Van Engelen, Inc.
23 Tulip Dr.
Bantam, CT 06750
(860) 567-8734
Bulbs

Andre Viette Farm and
  Nursery
State Rd. 608
Longmeadow Rd.
Fisherville, VA 22939
(540) 943-2315
Flowering shrubs, herbaceous perennials, ferns, grasses, Hemerocallis, *hostas, iris, peonies. Catalog: $5*

Washington Evergreen
  Nursery
PO Box 388
Leicester, NC 28748
(704) 683-4518
(April–October)
(803) 747-1641
(November–March)
Dwarf shrubs, rhododendrons, dwarf conifers, Kalmia. *Catalog: $2*

Waterford Gardens
74 East Allendale Rd.
Saddle River, NJ 07458
(201) 327-0721 or
327-0337
Aquatic plants, water lilies, lotuses. Catalog: $5

Waushara Gardens
North 5491 Fifth Dr.
Plainfield, WI 54966
(715) 335-4462
Gladiolus, *Asiatic lilies. Catalog: $1*

Wayside Gardens
1 Garden Lane
Hodges, SC
29695-0001
(800) 845-1124
Perennials, flowering
shrubs, trees

Wedge Nursery
Route 2, Box 114
Albert Lea, MN 56007
(507) 373-5225
Lilacs

We-Du Nurseries
Route 5, Box 724
Marion, NC 28752
(704) 738-8300
Southeastern native and
analogous Oriental wood-
land and rock plants; ferns,
species Iris, Trillium.
Catalog: $2

Well-Sweep Herb
  Farm
205 Mt. Bethel Rd.
Port Murray, NJ
07865
(908) 852-5390
Perennials, herbs, scented
geraniums. Catalog: $2

White Flower Farm
Route 63
Litchfield, CT
06759-0050
(800) 503-9624
Perennials, flowering shrubs,
roses, vines. Catalog: $5

White Rabbit Roses
PO Box 191
Elk, CA 95432
(707) 877-1888
Unusual roses

Wicklein's Water
  Gardens
PO Box 9780
Baldwin, MD 21013
(410) 823-1335
Aquatic and bog plants.
Catalog: $2

Gilbert H. Wild &
  Son, Inc.
PO Box 338
1112 Joplin St.
Sarcoxie, MO
64862-0338
(417) 548-3514
Hemerocallis, *iris,*
peonies. Catalog: $3

Nancy Wilson Species
  & Miniature Narcissus
6525 Briceland-Thorn
  Rd.
Garberville, CA 95542
(707) 923-2407
Species and miniature daf-
fodils. Catalog: $1

Windrose Nursery
1093 Mill Rd.
Pen Argyl, PA 18072
(610) 588-1037
Oaks, woody plants, flower-
ing shrubs, and ornamental
trees. Catalog: $3

Woodlanders, Inc.
1128 Colleton Ave.
Aiken, SC 29801
(803) 648-7522
Conifers, flowering shrubs,
trees, herbaceous plants,
ferns. Catalog: $2

Wrenwood of Berkeley
  Springs
Route 4, Box 361
Berkeley Springs, WV
25411
(304) 258-3071
Perennials, sedums, rock
plants. Catalog: $2.50

Guy Wrinkle Exotic
  Plants
11610 Addison St.
North Hollywood, CA
91601
(310) 670-8637
Cycads, caudiciforms, succu-
lents. Catalog: $1

Yucca Do Nursery
Rte. 3, Box 104
Hempstead, TX 77445
(409) 826-4580
Texas and southeastern
natives and conifers,
flowering shrubs.
Catalog: $4

Title page: bearded iris, white rock rose, California poppies, and grass. **Contents** page (clockwise from top left): Bellevue Botanical Garden border; *Drosanthemum speciosum; Fuschia austromontana* 'Autumnale', coleus, and white-edged ivy (*Hedra helix* 'Little Diamond'); unnamed selected seedling. **p. 8:** *Impatiens* sp. (*I. kew*). **p. 10:** *Clematis* and *Rodgersia pinnata* 'Superba'; purple Japanese iris and *Iris laevigata* 'Variegata'. **Woody Plants** chapter opener (clockwise from top left): lantana (*Lantana camara* 'Miss Huff'); *Enkianthus perulatus* 'Nana'; *Rubus fruticosa* 'Albavariegata'; *Aralia elata* 'Aurea Variegata'; *Xanthorhiza simplicissima;* cherry *Prunus serrula; Hydrangea macrophylla* 'Variegata'. **p. 16:** *Sambucus racemosa* 'Plumosa Aurea'. **Hardy Perennials** chapter opener (clockwise from top left): *Disporum smithii* 'Rick'; yellow slipper orchid (*Cypripedium calceeolus*); 'Sweet Harmony'; *Athyrium* x 'Branford Beauty'; *Hemerocallis* 'Little Vine Spider'; *Pulsatilla vulgaris; Arisaema ringens*. **p. 48:** *Primula aucaulis* 'Mark Viette'. **Tender Plants and Annuals** chapter opener (clockwise from top left): edible nasturtium flowers (*Tropaeolum nanum*); blood banana (*Musa zebrina*); Cleome, or spider flower (*Cleome spinosa* 'Helen Campbell'); variery of *Coleus* cultivars; *Elegeia capensis; Trichocereus schickendantzii.* **p. 80:** blanket flower (*Gaillardia pulchella*). **Climbing Plants** chapter opener (clockwise from top left): *Parthenocissus tricuspidata* 'Fenway'; old-fashioned cup and saucer vine (*Cobaea scandens*); *Mina lobata; Clematis* 'Betty Corning'; *Clematis florida* 'Sieboldii'; snail flower (*Vigna caracalla* [*Phaseolus caracalla*]); grape vine (*Vitis davidii*). **p. 100:** *Vitis vinifera* 'Purpurea'. **Appendix** opener (clockwise from top left): *Hemerocallis* 'Malaysia Monarch'; blue Himalayan poppy (*Meconopsis betonicifolia*); *Canna* 'Pretoria'; *Hemerocallis* 'Corky'; *Senecio mandraliscae; Asarum inflatum; Arisaema griffithii; Melianthus major;* fruits of the female *Osage* orange tree.

U.S. Plant Hardiness Zones: Approximate range of average annual minimum temperatures (°F): zone 1: below -50°; zone 2: -50° to -40°; zone 3: -40° to -30°; zone 4: -30° to -20°; zone 5: -20° to -10°; zone 6: -10° to 0°; zone 7: 0° to 10°; zone 8: 10° to 20°; zone 9: 20° to 30°: zone 10: 30° to 40°